Practical
Engineering Chemistry

Practical
Engineering Chemistry

Dr. K. Mukkanti
M.Sc., M.Phil, Ph.D. (Delhi)
Professor of Chemistry,
Institute of Science and Technology, (IST)
JNT University, Hyderabad - 50 0085, (A.P.).

BSP **BS Publications**
An Imprint of **BSP Books Pvt. Ltd.**
4-4-309/316, Giriraj Lane,
Sultan Bazar, Hyderabad - 500 095.

Practical **Engineering Chemistry** *by* **Dr. K. Mukkanti**

First Reprint 2024

Published by :

BSP BS Publications
An Imprint of **BSP Books Pvt. Ltd.**
4-4-309/316, Giriraj Lane, Sultan Bazar,
Hyderabad - 500 095
Phone : 040 - 23445688
e-mail : info@bspbooks.net
website: www.bspbooks.net

Printed at

Repro India Limited.

Price: Rs. 325.00

ISBN: 978-81-7800-201-9 (Paperback)

Dedicated

to my beloved Parents with immense love and regards

Late, Sri. Khagga Adinarayana

&

Smt. Khagga Adilakshmi

JAWAHARLAL NEHRU TECHNOLOGICAL UNIVERSITY HYDERABAD

Prof. DN. Reddy
B.E., M.Tech., Ph.D., FIE,MISME, MSESI, MIIPE, MISTE.
VICE-CHANCELLOR

Kukatpally
Hyderabd - 500 085
Andhra Pradesh (India)
Phone : 040-23156109 (O)
Fax : 040-23156112
E-mail : vcjntu@yahoo.com
WWW.jntuh.ac.in

FOREWORD

Since Chemistry is a rapidly growing subject, it is necessary to teach about the latest developments in this field. Chemistry is an experimental subject and this book describes the experimental foundations behind the scientific discoveries. Keeping in view in the recent developments in Engineering Chemistry, it has been written to have a better Knowledge of experimental chemistry to the engineering students. This book has been written in a simple language with in depth explanations, covers the principle and of engineering chemistry to cater the needs of the students.

The book entitled, "Practical Engineering Chemistry" Specially written for the engineering students will welcome by one and all. It satisfies the need of the first year engineering students of the JNTU and other Universities. It has been particularly written keeping in view of the objectives of the syllabus and requirements of the students. As there are only few books, which deal with practical approach for carrying out the experimental on Engineering Chemistry components like Using instrumental techniques such as Colorimetry, PHmetry, potentiometry, conductometry and flame photometry etc., have been progressively introduced. The main objective is to develop the analytical ability of the students by better understanding of the theory. A set of questions based on different experiments is given in the end of the book to make the students evaluate themselves.

The author Dr.K.Mukkanti, professor and Head, Centre for Pharmaceutical sciences, IST, JNTUH is a teacher, writer and Researcher. He knows the pulse of the students and teachers, who needs this information.

I hope this book will prove of real value to both the student community and the teachers.

PROF.DN.REDDY

PREFACE

Engineering technology has become inter-disciplinary with chemical, physical & mathematical sciences contributing immensely to the development of engineering practices. Any industrial process is strongly coupled with various observations on the basic properties and principles of the materials that they use. From Chemistry point of view, an engineering student is supposed to have a basic knowledge of properties like corrosion, phase rule, reactivity with surroundings, pH and other compatibility procedures. So the subject plays a vital role and to this end the present textbook on laboratory practices exclusively introduces concepts in a concise way, and is of great help.

- The book is written with the requirements of engineering students in mind. Every aspect of a topic is dealt with, keeping the focus on engineering science.
- The study of science is basically experimental and therefore, proper correlation between science theory and practicals shall assist in appreciation of the subject.
- The book has been written in a simple language with in-depth explanation.
- The viva questions at the end will help the students in developing a logical outlook.
- The book describes quantitative analysis and estimations of the compounds of general interest to engineers.
- It is hoped that this book will serve as a model for engineering students and the teachers of engineering chemistry.
- At the end, useful appendices including preparation of reagents are provided.
- As the accuracy of the results largely depends upon the accuracy of weights, volumetric glassware and solutions used, the first chapter introduces these features.
- This book includes with all relevant subject matter, equations and additional information.
- Instrumentation techniques such as pH meters, potentiometer, conductometer, and colorimeter and flame photometer are widely used in the industry and experiments covering these techniques have been introduced in this text
- The main feature of this book lies, in that it serves as a text on the practical engineering chemistry and provides both the theoretical and experimental approach to the subject in such simple manner that even an average student can perform the experimental work without much instruction.

Any constructive comments and suggestions from our valued readers for the further improvement of the book are solicited.

- Author

Acknowledgements

I take this opportunity to express deep love, gratitude and respect to my parents, (Late) Mr. Khagga Adinarayana and Mrs. K. Adilakshmi who were a driving force behind all my endeavors and will always remain so.

I sincerely thank Dr. D. N. Reddy, Vice-Chancellor, JNT University Hyderabad for writing the Foreword of my book.

I sincerely convey my appreciation and thanks to all my research students and colleagues Mr. K. Naresh Kumar, Mr. G.V. Siva Prasad, Mr. K. Kishore Kumar, Mrs. N. Mamatha and Mr. G. Dhilli Rao and who provided me the best possible assistance, during this prestigious work and helped me to accomplish the task to a great extent. I also thank my students Mr. P. Sadanandan, L. Venkata Reddy, and Y. Ramakoti Reddy for their help I am also thankful to all my friends and well wishers, who helped me throughout my life to achieve my goals in my career.

I am particularly thankful to Professor C. Parameswara Murthy, Department of Chemistry, Osmania University, Hyderabad for his encouragement and suggestions.

I am very much obliged to my family members – my wife **Nalini**, my beloved daughter, Bhavya Sri and son Sai Subhash for their inspiration and continuous support, without which it would have been impossible to complete this uphill task.

-Author

Contents

Precipitation Titrations

Analysis of Minerals

Cement Analysis

Ion Exchange Resin

Instrumentation

Chemical Kinetics

Viscosity, Surface Tension and Molecular Weight

Thermochemistry

Surface Tension

Electrochemistry

Part - III Organic Chemistry Experiments

Identification and Preparation of Organic Compounds

Part - I Inorganic Chemistry Experiments by Analytical Methods

EXPERIMENT – 1 Estimation of Hardness of Water by EDTA Method

INTRODUCTION

Water hardness is the traditional measure of the capacity of water to precipitate soap. Hard water requiring a considerable amount of soap to produce leather. Scaling of hot water pipes, boilers and other house hold appliances is due to hard water. Hardness of water is no specific constituent but is a variable and complex mixture of cations and anions. It is caused by dissolved polyvalent metallic ions. In fresh water, the principle hardness causing ions are calcium and magnesium. The other ions like Strontium, Iron, Barium and Manganese also contribute. Hardness is commonly expressed as $CaCO_3$ in mg/L. The degree of hardness of drinking water has been classified in terms of the equivalent $CaCO_3$ concentration as follows:

Soft	0-60 mg/L;
Medium	60-120 mg/L;
Hard	120-180 mg/L;
Very hard	> 180 mg/L;

Although hardness is caused by cation, it may also be discussed in terms of carbonate (temporary) and non-carbonate (permanent) hardness. Carbonate hardness refers to the amount of carbonates and bicarbonates in solution that can be removed or precipitated by boiling. This type of hardness is responsible for the deposition of scale in hot water pipes and kettles. When total hardness is numerically greater than that of total alkalinity expressed as $CaCO_3$, the amount of hardness equivalent to total alkalinity is called 'carbonate hardness'. The amount of hardness in excess of total alkalinity expressed as $CaCO_3$ is non-carbonate hardness. Non carbonate hardness is caused by the association of the hardness of causing cation with sulphate, chloride or nitrate and is referred to as "permanent hardness" because it can not be removed by boiling.

AIM

To estimate the amount of total hardness (Ca & Mg) present (as $CaCO_3$) in the given water sample by EDTA method.

APPARATUS

1. Conical flasks (100 mL)
2. Burette
3. Pipette
4. Spatula

CHEMICALS

1. Buffer solution
2. Inhibitor
3. Eriochrome black T indicator
4. Muroxide Indicator
5. NaOH (2N)
6. Standard EDTA Solution 0.01M
7. Standard Calcium Solution

THEORY

When Eriochrome Black T dye is added to the hard water at pH around 10 it gives wine red coloured unstable complex with Ca^{+2} and Mg^{+2} ions of the sample water. Now when this wine red-coloured complex is titrated agianst EDTA solution (of known strength) the colour of the complex changes wine red to original blue colour showing the endpoint.

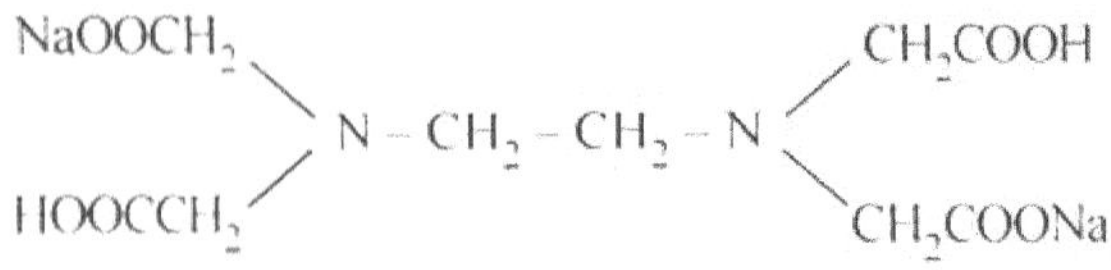

Disodium salt of ehtylenediamine tetraacetic acid: (Na_2H_2Y)

where Y = deprotonated agent.

In aqueous solution EDTA ionises to give $2Na^+$ ions and act as a strong chelating agent.

$$\left.\begin{matrix} Ca^{2+} \\ Mg^{2+} \end{matrix}\right\} + \text{Eriochrome black - T} \rightarrow \left[\begin{matrix} Ca^{2+} \\ \quad \text{Eriochrome black - T} \\ Mg^{2+} \end{matrix}\right] \text{Complex}$$

(of water) (Unstable complex) (Wine-red)

$$\xrightarrow{\text{EDTA}} \left[\begin{matrix} Ca^{2+} \\ \quad \text{EDTA} \\ Mg^{2+} \end{matrix}\right] \text{Complex} + \underset{\text{(blue)}}{\text{Eriochrome black - T}}$$

The indicator used is a complex organic compound (sodium – 1 – (1-hydroxy 2-naphthylazo)-6-nitro-2-naphthol-4-sutphonate). commonly known as Eriochrome black T(EBT). It has two ionisable phenolic hydrogen atoms and for simplicity it is represented as $Na^+H_2In^-$:

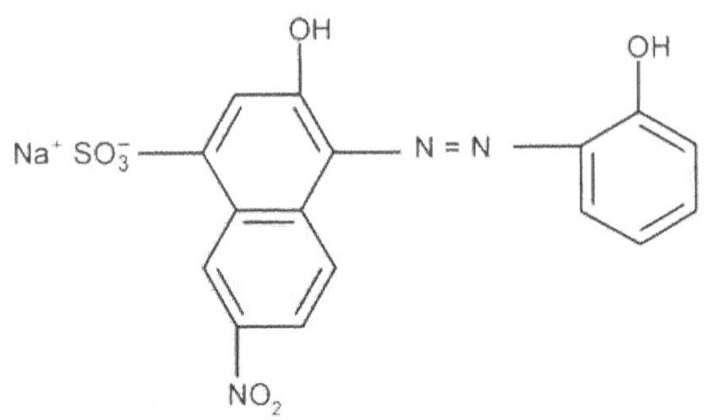

Eriochrome Black-T

Eriochrome Black-T(EBT) is the metal ion indicator used in the determination of hardness by complexometric titration with EDTA. This dye-stuff tends to polymerize in strongly acidic solutions to a red brown product, and hence the indicator is generally used in EDTA titration with solutions having pH greater than 6.5. The sulphuric acid gropus loses its proton much before the pH range of 7-12. which is of interest for metal ion indicator use. The dissociation of the two hydrogen atoms of the phenolic groups only should therefore be considered and hence the dye stuff may be represented by the formula

H_2D^-. This functions as acid-base indicator with two colour changes as follows:

$$\underset{\text{(Red)}}{H_2D^-} \underset{\text{pH 6.3}}{\rightleftharpoons} \underset{\text{(Blue)}}{HD^{2-}} \underset{\text{pH 11.5}}{\rightleftharpoons} \underset{\text{(Yellowish. Orange)}}{D^{3-}}$$

In the pH range 8-10. the blue form of the indicator HD^{2-} gives a wine red complex with Mg^{2+}:

$$Mg^{+2} + \underset{\text{(Blue)}}{HD^{2-}} \longrightarrow \underset{\text{(Wine red)}}{MgD^-} + H^+$$

Now if EDTA (H_2Y^{2-}) is added to such a solution Mg^{2+} preferentially complexes with EDTA (since the metal EDTA complex is more stable than the metal-indicator complex) and liberates the free indicator HD^{2-} at the end point. thereby producing a sharp colour change from wine red to blue. These reactions during the EDTA titration may be summarized as follows

$$Mg^{+2} + \underset{\text{(Blue)}}{HD^{2-}} \longrightarrow \underset{\text{(Wine red)}}{MgD^-} + H^+$$

$$H_2Y^{2-} + Ca^{2+} \longrightarrow CaY^{2-} + 2H^+$$

$$H_2Y^{2-} + Mg^{2+} \longrightarrow MgY^{2-} + 2H^+$$

$$H_2Y^{2-} + \underset{\text{(Winered)}}{MgD^-} \longrightarrow MgY^{2-} + \underset{\text{(Blue)}}{HD^{2-}} + H^+$$

The metal ion-indicators of common use in EDTA titrations include:

Eriochrome Black-T (or Solochrome Black). Muroxide (ammonium salt of purpuric acid). Eriochrome Blue-Black B (or Solochrome Black-6B). Patton and Reeders indicator. Solochrome Dark Blue or Calcon. Zincon. Xylenon Orange .... etc.

The optimum pH for the determination of hardness of water is 10.0 ± 0.1 and is adjusted by $NH_4OH - NH_4Cl$ buffer.

Advantages of EDTA method

This method is definitly preferable to the other methods. because of :

(i) Greater accuracy.

(ii) convenience and

(iii) more rapid procedure.

Units of Hardness

The followng units are used for expressing results.

1. *Parts per million (PPM)* : It is the number of parts of calcium carbonate equivalent hardness present in one million parts of water.
2. *Milligram per lite (mg/L)*: It is the number of milligrams of Calcium carbonate equivalent hardness present in one litre of water.
3. *Degree Clarke (°Cl)*: It is the number of parts of $CaCO_3$ equivalent hardness present in 70,000 parts of water.
4. *Degree French (°Fr)*: It is the number of parts of $CaCO_3$ equivalent hardness present in 10^5 (1 Lac) parts of water.

The above four units are correlated as given below

$$1PPM = 1mg/L = 0.07^{o}\ Cl = 0.1^{o}Fr$$

Determination of Hardness

The following of any given water sample may be determined by the following methods.

(i) O.Hehner's method

(ii) Soap titration method

(iii) EDTA method

EDTA Method

PREPARATION OF REAGENTS

1. ***Buffer solution:*** Dissolve 16.9 g NH_4Cl in 143 ml NH_4OH. Add 1.25 g magnesium salt of EDTA to obtain sharp change in colour of indicator and dilute to 250 ml. If magnesium salt of EDTA is not available, dissolve 1.179 g disodium salt of EDTA (AR grade) and 780 mg $MgSO_4.7H_2O$ or 644 mg $MgCl_2.6H_2O$ in 50 ml distilled water. Add this to above solution of NH_4Cl in NH_4OH and dilute to 250 ml.
2. ***Inhibitor:*** Dissolve 4.5 gm hydroxyl amine hydrochloride in 100 ml 95% ethyl alcohol or isopropyl alcohol.
3. ***Eriochrome black T(EBT) indicator:*** Mix 0.5 gm dye with 100 gm NaCl to prepare dry powder.
4. ***Muroxide Indicator:*** Prepare a ground mixture of 200 mg of murexide with 100 gm of solid NaCl.
5. ***NaOH (2N):*** Dissolve 80 gm NaOH and dilute to 1000 ml.
6. ***Standard EDTA Solution 0.01M:*** Dissolve 3.723 gm EDTA disodium salt and dilute to 1000 ml. Standardized against standard calcium solution, 1ml =1mg $CaCO_3$
7. ***Standard Calcium Solution:*** Weigh accurately 1gm $CaCO_3$ and transfer to 250 ml conical flask. Then add 1:1 HCl till $CaCO_3$ dissolve completely. Add 200 ml dist.water and boil for 20 to 30 min. then cool and add methyl red indicator. Add NH_4OH 3N drop wise till intermediate orange colour develops. Dilute to 1000 ml to obtain 1ml=1mg $CaCO_3$.

PROCEDURE

(a) Total hardness

(i) Take 25 or 50 ml or well mixed sample in a conical flask

(ii) Then add 1 to 2 ml buffer solution followed by 1 ml inhibitor

(iii) Add a pinch of Eriochrome black T and titrate with standard EDTA (0.01M) till **wine red** colour changes to **blue,** then note down the volume of EDTA required (**A**) .

(iv) Run a reagent blank. Note the volume of EDTA (**B**).

(v) Calculate volume of EDTA required by sample. **C** = **A** – **B** (from volume of EDTA required in steps (iii & iv).

(b) Calcium hardness

(i) Take 25 or 50 ml sample in a conical flask

(ii) Add 1 ml NaOH to raise pH to 12.0 and add a pinch of muroxide indicator.

(iii) Titrate immediately with EDTA till pink colour changes to **purple**. Note the volume of EDTA used (A_1).

(iv) Run a reagent blank. Note the ml of EDTA required (B_1) and keep it aside to compare end points of sample titrations.

(v) Calculate the volume of EDTA required by sample, $C_1 = A_1 - B_1$.

OBSERVATIONS AND CALCULATIONS

Water Sample Vs EDTA

S. No	Volume of water sample (ml)	Burette Reading		Volume of EDTA added (ml)
		Initial (ml)	Final (ml)	
1.				
2.				
3.				
4.				

(i) Total hardness as $CaCO_3$, mg /l = $\dfrac{C \times D \times 1000}{\text{Volume of sample in ml}}$

Where

C = Volume of EDTA required by sample (with EBT indicator)

D = mg $CaCO_3$ equivalent to 1 ml EDTA titrant (1 ml 0.01 MEDTA ≡ 1.000 mg $CaCO_3$)

or ($D = 1 \times \dfrac{\text{Molarity of EDTA}}{0.01M}$)

(ii) Calcium hardness as $CaCO_3$, mg /l = $\dfrac{C_1 \times D \times 1000}{\text{Volume of sample in ml}}$

Where

C_1 = volume of EDTA used by sample (with murexide indicator)

D = mg $CaCO_3$ equivalent to 1 ml EDTA titrant

(iii) Magnesium hardness

Magnesium Hardness = Total hardness as $CaCO_3$, mg/l – Calcium hardness as $CaCO_3$, mg/l.

RESULT

Amount of total hardness present in the given water sample = ——— mg/l

EXPERIMENT – 2 Estimation of Alkalinity of Water

INTRODUCTION

Alkalinity of water is the measure of its capacity to neutralize acids. Alkalinity of natural water is due to the salts of carbonates, bicarbonates, borates, silicates and phosphates along with hydroxyl ion in the free state. However, the major portion of alkalinity in natural water is caused by hydroxides, carbonates and bicarbonates which may be ranked in order of their association with high pH values.

Alkalinity is primarily due to the presence of carbonate, bicarbonate and hydroxide ions in natural waters. The bicarbonates of calcium and magnesium results from the reaction of the acidic water with calcium and magnesium minerals such as *calcite* ($CaCO_3$). Dolomite ($MgCa(CO_3)_2$ and *Gypsum* ($CaSO_4 \cdot 2H_2O$).

$$H_3O^+ (aq) + CaCO_3 (s) \rightleftharpoons Ca^{2+}(aq) + HCO_3^- (aq) + H_2O$$

The alkalinity of water is a measure of its acid neutralising capacity and is usually related to the concentration of carbonate, bicarbonate and hydroxide. Low alkalinities (25 g/m^3) promote corrosion and high alkalinities can cause problems with scale formation in metal pipes, tanks and heating systems. ***Total alkalinity is the total concentration of bases in water expressed as parts per million (ppm) or milligrams per liter (mg/l) of calcium carbonate ($CaCO_3$).***

pH range of 6 to 8 is found to be the best for most fish and insects in streams. Fish, insects, and bacteria are able to survive at pH higher and lower than this, but not for long. Bacteria (insect larvae feed on these) begin to die off below a pH of 5.5. Insects and other macroinvertebrates begin to die off at a pH of 5 (these feed fish). Fish are mostly gone at pH of 4.5 or lower. So, slight alkalinity of natural waters is a requirement for aquatic life.

The most common method of increasing alkalinity in water is by adding agricultural limestone (calcium carbonate). Fish farms in Europe and Asia have reported that applications of lime to fish ponds on soils of low calcium content resulted in greater fish production.

In the first experiment you will determine the total alkalinity of water. This is done by titrating a known quantity of natural water against standard hydrochloric acid. The indicator used is methyl orange.

AIM

To determine the amount of alkalinity in the supplied water sample.

APPARATUS

1. Beakers
2. Pipettes
3. Volumetric flasks (1000 ml, 200 ml, 100 ml)

CHEMICALS

1. 0.02 N Sulphuric acid (H_2SO_4)
2. Phenolphthalein indicator
3. Methyl orange indicator

THEORY

Alkalinity of sample can be estimated by titrating with standard sulphuric acid titration to pH 8.3 or decolourization of phenaphthalein will indicate the partial neutralization of hydroxides and half of carbonates, while to pH 4.5 or sharp colour change from yellow to orange of methyl orange indicator will indicate total alkalinity (complete neutralization of the bicarbonate ions).

Alkalinity in natural waters is due to free hydrolysis and hydrolysis of salts formed by weak acids and strong bases.

$$A^- + HOH \longrightarrow HA + OH^-$$

Most of the alkalinity in natural water is formed due to dissolution of CO_2 in water. Carbonates and bicarbonates thus formed are dissolved to yield hydroxyl ions.

$$CO_2 + H_2O \rightleftharpoons H_2CO_3$$

$$H_2CO_3 \rightleftharpoons H^+ + HCO_3^-$$

$$HCO_3^- \rightleftharpoons H^+ + CO_3^{2-}$$

$$CO_3^{2-} + 2H_2O \rightleftharpoons H_2CO_3 + 2OH^-$$

$$HCO_3^- + H_2O \rightleftharpoons H_2CO_3 + OH^-$$

Therefore, the total alkalinity (TA) of the system is due to

$$TA = HCO_3^- + CO_3^{2-} + OH^-$$

Thus, the alkalinity of water is mainly due to hydroxides, carbonates and bicarbonates. The following possibilities may arise with respect to the constituents causing alkalinity in natural waters:

(i) Hydroxides (ii) Carbonates (iii) Bicarbonates

(iv) Hydroxide and carbonates (v) Carbonates and bicarbonates.

Hydroxides and bicarbonates cannot exist together as they combine with each other to form carbonates.

$$OH^- + HCO_3^- \longrightarrow CO_3^{2-} + OH^-$$

The type and extent of alkalinity present in a water sample can be determined by titrating the sample with a standard solution of acid using phenolphthalein (P) (1 & 2 equations) and methyl orange (M) (1, 2 & 3 equations) indicators.

$$Ca(OH)_2 + H_2SO_4 \longrightarrow CaSO_4 + 2H_2O \qquad (1)$$

$$2CaCO_3 + H_2SO_4 \longrightarrow Ca(HCO_3)_2 + CaSO_4 \qquad (2)$$

$$Ca(HCO_3)_2 + H_2SO_4 \longrightarrow CaSO_4 + 2CO_2\uparrow + 2H_2O \qquad (3)$$

The volume of acid run-down for phenolphthalein (phe) end point, P *ml*, corresponds to the completion of equations (1) and (2) while the volume of acid run-down after P *ml* corresponds to the completion of equation (3).

The total amount of acid used from the beginning of the experiment, i.e., (M) corresponds to the total alkalinity and represents the completion of reaction shown by equations (1) to (3).

The titration values obtained from phenolphthalein and total alkalinity determination are used to calculate the three forms of alkalinity as shown in the Table:

Phe end Point (P) and MeO end Point (M)	OH^- Alkalinity as $CaCO_3$	CO_3^{2-} Alkalinity as $CaCO_3$	HCO_3^- Alkalinity as $CaCO_3$
P = 0	0	0	M
$P < \frac{1}{2} M$	0	2P	M – 2P
$P = \frac{1}{2} M$	0	2P or M	0
$P > \frac{1}{2} M$	2P – M	2(M – P)	0
P = M	P or M	0	0

Where P is phenolphthalein alkalinity and M is total alkalinity.

Alkalinity in itself is not harmful to human beings, but water supplies with less than 100 mg/*l* are desirable for domestic use. Also, highly alkaline water may lead to caustic embrittlement and may cause deposition of precipitates and sludges in boiler tubes. Alkali measurements are also important in controlling water and wastewater treatment process.

PREPARATION OF REAGENTS

1. ***0.02 N Sulphuric acid* (H_2SO_4):** Prepare 0.1 N H_2SO_4 by diluting 3 ml conc.H_2SO_4 to 1000 ml. Standardize it against standard 0.1 N Na_2CO_3 by using methyl orange as indicator. Dilute appropriate volume of H_2SO_4 to 1000 ml to obtain standard 0.02 N H_2SO_4.
2. ***Phenolphthalein indicator*:** Dissolve 0.5 g in 500 ml of 95% ethyl alcohol. Add 500 ml distilled water. Add drop wise 0.02 N NaOH till faint pink colour appears (pH 8.3).
3. ***Methyl orange indicator*:** Dissolve 0.5 g and dilute to 1000 ml with CO_2 free distilled water (pH 4.3-4.5).

PROCEDURE

(i) Take 25 or 50 ml sample in a conical flask and add 2-3 drops of phenolphthalein indicator.

(ii) If pink colour develops titrate with 0.02N H_2SO_4 till it disappears or pH is 8.3. Note the volume of H_2SO_4 required (**A**).

(iii) Add 2-3 drops of methyl orange to the same flask, and continue titration till pH comes down to 4.5 or orange colour changes to pink. Note the volume of H_2SO_4 added (**B**).

(iv) In case pink color does not appear after addition of phenolphthalein continue as in step 3 above.

(v) Repeat the experiment until to get the concurrent readings.

OBSERVATIONS AND CALCULATIONS

Water Sample Vs H_2SO_4

S.No	Volume of sample(ml)	Burette reading		Volume of H_2SO_4 consumed (ml)
		Initial reading (ml)	Final reading (ml)	
1.				
2.				
3.				

Calculate total (T), phenolphthalein (P) alkalinity as follows:

P- Alkalinity, (mg/l as $CaCO_3$) = A × 1000 / Volume of sample (ml)

T - Alkalinity, (mg/l as $CaCO_3$) = B × 1000 / Volume of sample (ml)

In case H_2SO_4 is not 0.02N, apply the following formula:

$$\text{Alkalinity (mg/l as } CaCO_3) = \frac{\textbf{A or B} \times N \times 50000}{\text{Volume of sample (ml)}}$$

where

A = ml of H_2SO_4 required to bring the pH to 8.3 (with phenolphthalein indicator)

B = ml of H_2SO_4 required to bring the pH to 4.5 (with methyl orange indicator)

N = normality of H_2SO_4

Once, the phenolphthalein and total alkalinities are determined, three types of alkalinities, i.e. hydroxide, carbonate and bicarbonate are easily calculated.

The conversions are as follows:

mg CO_3^{-2} / L = Carbonate alkalinity mg $CaCO_3$/L × 0.6

mg HCO_3^-/L = Bicarbonate alkalinity mg $CaCO_3$/L × 1.22

RESULT

The amount of total alkalinity present in the water sample is ——— mg/L

EXPERIMENT – 3

Determination Percentage of Available Chlorine in the given Sample of Bleaching Powder

INTRODUCTION

Bleaching powder, which is a mixture of calcium hypochlorite $CaOCl_2$ basic calcium chloride $CaCl_2.Ca(OH)_2.2H_2O$ and a little slaked lime, but its main constituent calcium hypochlorite is mainly responsible for the bleaching action.

$$OCl^- + Cl^- + 2H^+ \rightarrow Cl_2 + H_2O$$

On reaction of dilute acids with bleaching powder, chlorine is liberated which is also known as "available chlorine". The amount of chlorine liberated is also a method for determining the quality of bleaching powder. A good sample of bleaching powder usually contains 35-39% of available chlorine.

AIM

To determine the percentage of available chlorine in the given sample of bleaching powder.

APPARATUS

(a) Burette (b) Pipette (c) Volumetric flask (d) Measuring Jar
(e) Burette stand (f) Watch glass (g) Weighing tube

CHEMICALS

1. 0.05 N Copper sulphate solution
2. 0.05 Sodium thiosulphate solution
3. Starch indicator (1%)
4. Potassium iodide solution (25%)

THEORY

In the present method discussed below, chlorine liberated by the action of dilute acid on bleaching powder replaces iodine from potassium iodide solution. Since the liberated iodine is equivalent to the chlorine present in the bleaching powder, hence iodine can be estimated by titrating it with standard sodium thiosulphate solution.

$$Cl_2 + 2KI \rightarrow I_2 + 2KCl$$

$$I_2 + 2Na_2S_2O_3 \rightarrow 2NaI + Na_2S_4O_6$$

PREPARATION OF SOLUTIONS

1. ***Standard N/20 copper sulphate (Eq.wt. 249.7) solution:*** Weigh accurately (249.7/20 × 4 =3.121) 3.2 grams of copper sulphate and transfer it in a 250 ml measuring flask containing 50 ml of distilled water and 5 ml of dilute acetic acid. Stopper the flask and shake thoroughly so that copper sulphate completely dissolves. Now make up the volume up to 250 ml mark by adding distilled water.
2. ***N/20 Sodium thiosulphate(Eq.wt. 248.20)solution:*** A standard solution of sodium thiosulphate cannot be prepared by direct weighing because it decomposes in solution due to bacterial action.

sulphur gets deposited and as a result the strength of solution gets changed. However, a solution of approximate N/20 normality is prepared by dissolving (248.2/10 x 4 = 6.4) 6.4 grams of sodium thiosulphate in 250 ml of distilled water. This solution is kept overnight to make a homogeneous.

3. ***Starch indicator(1%):*** Make a paste of 1 gm of soluble starch in 5-7 ml of water and pour the paste with constant stirring in 100 ml of boiling water. Allow the solution to cool to room temperature.
4. ***Potassium iodide solution(25%):*** Dissolve 25 gms of potassium iodide in 100 ml of distilled water.

PROCEDURE

Standardization of Sodium thiosulphate solution:

(i) Fill the clean dry burette with sodium thiosulphate solution.

(ii) Pipette out 25 ml of copper sulphate solution in a 100 ml conical flask and then add 5 ml of 25% potassium iodide solution to it.

(iii) Cover the mouth with watch glass and then keep it in some dark place for 2-3 minutes. The copper sulphate solution turns brownish in color due to liberation of iodide. Place a white paper below the conical flask and run the sodium thiosulphate solution dropwise and with constant shaking till the solution in the conical flask assumes light or pale yellow colour.

(iv) Add 1 ml of starch solution to this yellow coloured solution. The solution in the conical flask immediately turns blue.

(v) Continue the drop wise addition of the sodium thiosulphate solution from burette until the blue colour just disappears. This is the end point. Note the volume of sodium thiosulphate used.

(vi) Repeat the titration process in order to get two concordant readings.

OBSERVATIONS AND CALCULATIONS

Part - 1

Weight of empty weighing tube = m gms

Weight of weighing tube + copper sulphate = m_1 gms

Weight of copper sulphate = $(m_1 - m)$ gms

Normality of copper sulphate (Eq. wt. 249.7) = $4 \times (m_1 - m)/249.7$

$CuSO_4$ Solution Vs $Na_2S_2O_3$ Solution

S.No	Volume of copper sulphate solution taken (ml)	Volume of sodium thiosulphate used (ml)
1	25	Say V_1 ml
2	25	Say V_1 ml
3	25	Say V_1 ml

Part - 2

$CaOCl_2$ Solution Vs $Na_2S_2O_3$ Solution

S.No	Volume of bleaching powder solution taken (ml)	Volume of sodium thiosulphate used (ml)
1	25	Say V_2 ml
2	25	Say V_2 ml
3	25	Say V_2 ml

Volume × Normality (of bleaching powder) = Volume × Normality of ($Na_2S_2O_3$)

25 × Normality of (bleaching solution) = $V_2 \times 25 \times 4(m_1 - m)/V_1 \times 249.7$

Normality of sodium thiosulphate = $V_2 \times 4 \times (m_1 - m)/V_1 \times 249.7$

$$\text{Normality of the available chlorine in gm/litre} = \frac{V_2 \times 4 \times (m_1 - m)}{V_1 \times 249.7} \times 35.5$$

$$\text{Percentage of available chlorine} = \frac{V_2 \times 4 \times (m_1 - m)}{V_1 \times 249.7} \times 35.5 \times \frac{250}{1000} \times \frac{100}{w}$$

$$\frac{V_2 \times 4 \times (m_1 - m) \times 35.5 \times 25}{V_1 \times 249.7 \times w}$$

PRECAUTIONS

1. The glass apparatus used in the experiment should be washed with distilled water.
2. The volume of indicator should be same in all the titrations.
3. All the reagents should be freshly prepared.
4. End point of the titration should be carefully observed.

RESULT

The percentage of chlorine in given sample of bleaching powder = ——————

EXPERIMENT – 4 Estimation of Dissolved Oxygen in Water

INTRODUCTION

Determination of Dissolved Oxygen (D.O.) is important for industrial purposes. Dissolved Oxygen is needed for living organism to maintain their biological processes. D.O. is also important in precipitation and dissolution of inorganic substances in water. Dissolved Oxygen is an important factor in corrosion. D.O. test is used to control the amount of oxygen in boiler feed water by mechanical, physical and chemical methods. This test helps to assess raw water quality and to keep a check on stream pollution. D.O. test is the basis of B.O.D. (Biological test which is an important parameter in evaluating the pollution potential of domestic wastes).

Oxygen is poorely soluble in water. The solubility of D.O.. decreases with increase in conc. of the salt under a pressure of one atmosphere. The solubility of oxygen of air in distilled or fresh waters with low solid concentrations varies from 14.5 mg/L, at 0°C to about 7.5 mg/*l* at 30°C. The solubility is less in saline waters and at a given temperature decreases with increase in the concentration of impurities. Iodometric (Winkler's method) and electrometric methods using electrode are the two methods used for determining D.O. in water.

AIM

Determination of Dissolved Oxygen present in given water sample.

APPARATUS

1. Pipette
2. Burette
3. Glass bottles
4. Glass rod
5. Conical flask
6. Measuring Jar

CHEMICALS

1. Standard sodium thio sulphate solution (N/50)
2. Potassium permanganate solution (N/10)
3. Potassium oxalate solution (2%)
4. Manganous sulphate solution (4.8%)
5. Alkaline potassium Iodide
6. Freshly prepared starch solution
7. Concentrated sulphuric acid

THEORY

The principle involved in the determination of D.O. is to bring about the oxidation of potassium iodide to iodine with the dissolved oxygen present in the water sample after adding $MnSO_4$. KOH & KI. the basic manganic oxide formed acts as an oxygen carrier to enable the dissolved oxygen in the molecular form to take part in the reaction. The liberated iodine is titrated against standard sodium thiosulphate (HYPO) solution, using starch as indicator.

$$MnSO_4 + 2KOH \longrightarrow \underset{\text{(white)}}{Mn(OH)_2} + K_2SO_4$$

$$2Mn(OH)_2 + O_2 \longrightarrow \underset{\text{(brown ppt)}}{2MnO(OH)_2} \text{ Basic manganic oxide}$$

$$MnO(OH)_2 + H_2SO_4 \longrightarrow MnSO_4 + 2H_2O + [O]$$

$$2KI + H_2SO_4 + [O] \longrightarrow K_2SO_4 + H_2O + I_2$$

$$I_2 + 2Na_2S_2O_3 \longrightarrow Na_2S_4O_6 + 2NaI$$

$$\text{Starch} + I_2 \longrightarrow \text{Blue coloured complex.}$$

PREPARATION OF REAGENTS

1. ***0.02N $Na_2S_2O_3$:*** Dissolve 4.9642 g of $Na_2S_2O_3 .5H_2O$ (Mol. Wt. 248.21: eq. wt. 248.21) in 1 litre boiled out distilled water. Preserve by adding 0.2 g solid NaOH or 1 ml of 6 N NaOH and dilute to 1000 ml.
2. ***Potassium permanganate solution (N/10):*** Dissolve 3.1607 g of $KMnO_4$ having molecular weight 158.037 and equivalent weight 31.607 in distilled water. Boil for 1 hour, cool, filter and make up the volume to 1 litre with distilled water.
3. ***Alkaline KI solution:*** Dissolve 10 g KOH and 5 g of KI in 20 cc of previously boiled distilled water and filtered.
4. ***Manganese sulphate :*** Dissolve 10 g $MnSO_4 .4H_2O$ in 20 ml of boiled distilled water and add few drops of formaldehyde.
5. ***1% starch solution:*** Make a paste of 1 g of soluble starch in 5-7 ml of water and pour the paste with constant stirring in 100 ml of boiling water. Allow the solution to cool to room temperature.
6. **Conc. H_2SO_4** (specific gravity is 1.84)

PROCEDURE

(i) Take N/50 hyposolution in a burette and fix in the burette stand.

(ii) Collect the water sample in a 300 ml glass stoppered bottle and with the help of a graduated pipette, add 0.9 ml conc. H_2SO_4 and 0.2 ml (4 drops) $KMnO_4$ solution. Stopper the bottle and mix the contents of the bottle by inverting it a few times. If the permanganate colour disappears within 5 minutes, add additional amount of $KMnO_4$.

(iii) Add 0.5 ml of potassium oxalate solution (2%), stopper and mix well. Add additional amount of oxalate solution if the permanganate colour is not discharged within 10 minutes.

(iv) Add 2 ml of $MnSO_4$ solution followed 3 ml of alkaline KI solution. Stopper and shake and allow the precipitate to settle.

(v) Add 1 ml of concentrated H_2SO_4 solution and mix until the precipitate is completely dissolved. Measure 102.2 ml of this solution with a measuring cylinder in to a conical flask and titrate slowly with hypo. When the colour of the solution is very light yellowish add about 2 ml of freshly prepared starch solution and continue the titration to the disapperance of the blue colour and note down the volume of hypo used.

(vi) Repeat the titration to get the concordant volume of hypo used.

OBSERVATIONS

Total volume of the sample taken = 300 ml

Volume of reagents added during the preparation of iodine solution

$$= 0.9 \text{ ml } H_2SO_4 + 0.2 \text{ ml } KMnO_4 + 0.5 \text{ ml } K_2C_2O_4 + 2 \text{ ml } MnSO_4 + 3 \text{ ml alkaline KI}$$

$$= 6.6 \text{ ml.}$$

Volume of prepared solution (Iodine) taken for titration = 102.2 ml

Concordant volume of N/50 Hypo solution used = V ml.

CALCULATIONS

1 ml of 1N $Na_2S_2O_3$ = 8mg of O_2.

6.6 ml of the reagents have been added under such conditions that approximately equal volume of the sample is displaced. This dilutes the sample and so a correction is needed.

N (Normality of the sample with respect to D.O.)

$$= \frac{\text{Hypo normality}\left(\frac{N}{50}\right) \times \text{Volume of Hypo (V)}}{\text{Volume of the original sample equivalent to 102.2 ml (100 ml)}}$$

$$= \frac{1}{100} \times \frac{1}{50} \times V \times 8 \text{ g/}l$$

$$= \frac{1}{100} \times \frac{1}{50} \times V \times 8 \times 1000 \text{ mg/}l$$

$$= 1.6 \text{ V mg/}l$$

(or) $$\text{D.O.(mg/}l) = \frac{V \times 0.16 \times 1000}{\text{Volume of sample}}$$

Volume of the original sample that will be equivalent to 102.2 ml of the prepared (Diluted) solution is

$$= \frac{\text{volume of the sample} - \text{volume displaced by reagents.}}{\text{Total volume of the sample}} \times 102.2$$

$$= \frac{(300 - 6.6) \times 102.2}{300} = 100 \text{ ml}$$

RESULT

Dissolved oxygen preset in the given water sample = ________ (mg/*l*).

Note

(i) D.O. in boiler water is responsible for boiler corrosion, which can be controlled by removing D.O. from the water by heating it under pressure and by adding calculated quantity of Na_2S, Na_2SO_3, hydrazine (N_2H_4) to the boiler water.

(ii) Dissolved oxygen in water is also necessary for aerobic biological activities. In the absence of sufficient amount of dissolved oxygen in water, the anaerobic degradation of the pollutants make the water foul smelling. D.O test is helpful in determining the pollution extent of sewage or any other polutant.

EXPERIMENT - 5 Preparation of a Nickel Complex $[Ni(NH_3)_6]Cl_2$ and Estimation of Nickel by Complexometric Titration

INTRODUCTION

In this experiment, preparation and analysis of a simple Ni(II) complex, $[Ni(NH_3)_6]Cl_2$ will be carried out.

In the reactions of nickel chloride hexahydrate with ammonia, it has been known that initially there can be formation of nickel hydroxide (observed as a green precipitate), which reacts further with ammonia to give ammine complexes. It has also been shown by separate experiments that nickel hydroxide reacts with ammonia to give ammine complexes of Ni(II). It may also be noted that the reaction is dependent upon the stoichiometry and if excess of ammonia is not used, one can end up getting a mixture of products with varying compositions (e.g. $[Ni(NH_3)_5(H_2O)]Cl_2$, $[Ni(NH_3)_3(H_2O)_3]Cl_2$, $[Ni(NH_3)_5Cl]Cl$ etc.

To make sure that the right complex has been made, one has to estimate the amount of nickel, ammonia and chloride in the complex. (Ammonia can be estimated by acid base titration and chloride can be estimated gravimetrically by converting it to AgCl with $AgNO_3$. These are however, not taken up in this experiment). Nickel can be estimated by three well known methods:

(a) by spectrophotometric analysis (wherein the ammine complex is converted back to the aqua complex and its absorbance compared with standard solutions of nickel at a specific wavelength of ~395 nm).

(b) by gravimetric analysis after converting nickel to its bis(dimethylglyoximato) complex and

(c) by complexometric titration with EDTA.

Preparation of the $[Ni(NH_3)_6]^{2+}$ complex

The complex is prepared by taking nickel chloride hexahydrate as a water solution to which an excess of aqueous ammonia is added. The colour of the solution changes from pale dark green to purple. The solution is then cooled in an ice bath upon which the $[Ni(NH_3)_6]^{2+}$ complex will be precipitated out as purple crystals. This can be separated by filtration, washed with ammonia solution and dried.

$$NiCl_2 \cdot 6H_2O + 6NH_3 \rightarrow Ni(NH_3)_6 Cl_2 + 6H_2O$$

In this experiment, you will also estimate the amount of nickel present in the complex by complexometric titration using the well known chelating ligand EDTA. EDTA reacts with metal ions to form very stable complexes. In many complexes EDTA completely engulfs the metal ion, forming a six coordinate species as shown above. Two nitrogen atoms occupy adjacent positions of the octahedrally coordinated metal ion.

EDTA titrations involve the use of an indicator such as Eriochrome black T or Murexide. These are metal ion indicators which are compounds whose colour changes when it binds to a metal ion. For an indicator to be useful it must bind metal less strongly than EDTA does. The complex is first converted to

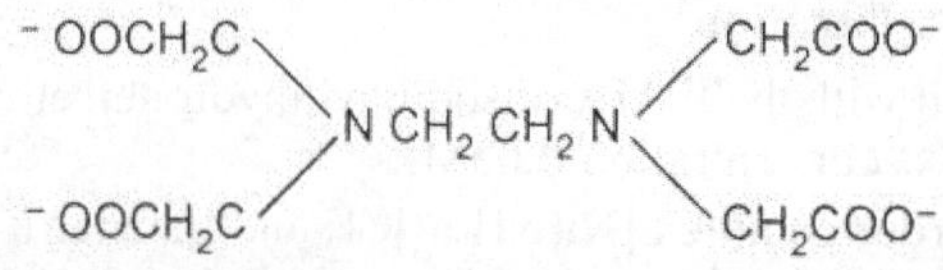

nickel sulfate. The added murexide indicator forms a weak complex with the metal solution attaining yellowish colour during titration. The titration is to be carried out at pH 7 for a sharp end point. At the end point, the colour of the free indicator is seen which is bluish violet. pH 7 is maintained by adding a buffer composed of ammonia and ammonium chloride. Ammonia, a weak base raises the pH by forming hydroxide ions. It also neutralizes added acids to prevent lowering of pH while the ammonium ion can react with any added hydroxide to form ammonia and water thus preventing variations in the pH.

$$\mathbf{Ni(NH_3)_6\,Cl_2 + EDTA^{-4} \rightarrow Ni(EDTA)^{-2} + 6NH_3 + 2Cl^-}$$

$$\mathbf{Ni^{2+} + EDTA^{-4} \rightarrow NiEDTA^{-2}}$$

$$\mathbf{Ni\,(mur)}\,\text{(yellow)} + \mathbf{EDTA^{-4} \rightarrow NiEDTA^{-2} + Mur^{-2}}\,\text{(bluish violet)}$$

The colour change occurs over a range of 2 drops of EDTA and turns from yellow to bluish violet.

AIM

To prepare nickel complex and estimate the nickel by complexometric titration.

APPARATUS

1. Beakers 2. Glass rod 3. Funnel 4. Burette 5. Pipette 6. Standard flask 7. Conical flask

CHEMICALS

1. Nickel chloride hexahydrate 2. Conc. aqeous ammonia 3. EDTA 4. 1N H_2SO_4 5. 0.5M ammonium chloride 6. Murexide indicator.

PROCEDURE

I. Preparation of Hexamine nickel (II) chloride

(i) Take 5 ml of a solution of Nickel chloride hexahydrate (containing about 6 g of nickel chloride) in a 100 ml beaker.

(ii) Take 10 ml of conc. aqueous ammonia in another beaker.

(iii) Add the aqueous ammonia slowly to a rapidly stirred solution of the nickel chloride.

(iv) Make sure that the colour of the solution has changed from pale green to intense purple (if not, add more of ammonia).

(v) Allow the solution to stand at room temperature for 1-2 mins, cover it with filter paper or aluminum foil and then cool it in a crushed ice bath without disturbance for about 15 mins.

(vi) If purple-blue crystals of $[Ni(NH_3)_6]Cl_2$ has not formed, try initiating crystallization by gently scratching the sides of the beaker with a glass rod.

(vii) Decant carefully the solution or filter it to separate the crystals

(viii) Dry the crystals by keeping it between a few pieces of filter paper.

(ix) Determine the weight of the dried complex and calculate the percentage yield.

II. Estimation of Nickel by EDTA Titration

(i) Clean your burette and fill it with the EDTA (disodium dihydrogen ethylenediamine tetra acetate dehydrate) solution (approx. concentration 0.05M).

(ii) Weigh accurately 0.6 g of your complex $[Ni(NH_3)_6]Cl_2$ and transfer it to a 50 ml standard flask. add 10 ml of 1N sulfuric acid. Dissolve and make up the solution to the mark.

(iii) Pipette out 10 ml of the solution of the complex into a 250 ml conical flask.

(iv) Dilute the solution with 15 ml of distill water.

(v) Add 4-5 drops of freshly prepared murexide indicator to the conical flask.

(vi) Add approximately 5 ml of 0.5M Ammonium chloride solution.

(vii) Add a few drops ammonia solution until the pH is about 7 as indicated by the yellow colour of the solution (this step is not required if the solution is already yellow).

(viii) Titrate with EDTA till the end point is approached which is indicated by the yellow solution turning to pale brown (this darkening may sometimes be seen only after addition of ammonia).

(ix) Add about 3 ml of ammonia solution. continue titration till the end point at which colour changes from yellow to bluish violet.

(x) Repeat till you get concordant values and calculate the strength of the nickel solution. and from that the amount of nickel present in the complex.

(xi) Compare with the expected percentage of nickel in the weighed sample of complex and determine the error if any in the estimation.

OBSERVATIONS AND CALCULATIONS

Ni Complex Solution Vs EDTA Solution

S.NO.	Volume of Ni complexes solution	Burette Reading		Volume of EDTA (x ml)
		Initial (ml)	Final (ml)	
1	10 ml			
2.	10 ml			
3.	10 ml			

1 ml 0.1M EDTA = 5.871 mg of Ni

1 ml 0.05M EDTA = 2.935 mg of Ni

1 ml 0.05M EDTA = 2.935 × x ml of Ni

$$\% \text{ of yield} = \frac{\text{Practical yield}}{\text{Theoritical yield}} \times 100$$

Amount of Ni present in given (0.6g in 50 ml) complex solution) = 2.935 × x × 5 mg

= ——— gm

RESULT

% yield of the complex = ——— %

Amount of Nickel present in given complex solution = ——— gm

EXPERIMENT – 6 Determination of Aluminium by Back Titration Method

INTRODUCTION

A back-titration is a laboratory operation where you titrate indirectly. Rather than titrate the analyte directly, you react the analyte with an intermediary. Then, you titrate the excess intermediary to determine how much was left over. The difference between how much intermediary you added and how much is left over is equivalent to the amount of analyte.

There are at least three reasons: (1) to convert an unstable reagent to another form and preserve the integrity of the sample. (2) to change the direction of the end point colour change, or (3) to convert a material to a more convenient form.

AIM

To estimate the amount of aluminium present in the given sample by back titration method.

APPARATUS

1. Conical flasks 2. Burette 3. Pipette

CHEMICALS

1. Buffer solution
2. Eriochrome black T indicator
3. Murexide Indicator
4. Standard EDTA Solution 0.01M
5. Standard $ZnSO_4$ Solution 0.01M
6. Dil. Ammonia

THEORY

Many metals cannot, for various reasons, be titrated directly, thus they may precipitate from the solution in the pH range necessary for the titration, or they may form complexes too slowly, or a suitable metal indicator is not available. In such cases an excess of standard EDTA solution is added, the resulting solution is buffered to the desired pH, and the excess of the reagent is back titrated with a standard metal ion solution. A solution of zinc chlorid or sulphate or of magnesium chloride or sulphate is often used for this purpose. The end point is detected with the aid of the metal indicator which responds to the metal ion introduced in the back titration.

PREPARATION OF REAGENTS

1. ***Buffer solution:*** Dissolve 16.9 g NH_4Cl in 143 ml NH_4OH. Add 1.25 g magnesium salt of EDTA to obtain sharp change in colour of indicator and dilute to 250 ml. If magnesium salt of EDTA is not available, dissolve 1.179 g disodium salt of EDTA (AR grade) and 780 mg $MgSO_4.7H_2O$ or 644 mg $MgCl_2.6H_2O$ in 50 ml distilled water. Add this to above solution of NH_4Cl in NH_4OH and dilute to 250 ml.
2. ***Eriochrome black T indicator:*** Mix 0.5 gm dye with 100 gm NaCl to prepare dry powder.
3. ***Murexide Indicator:*** Prepare a ground mixture of 200 mg of murexide with 100 gm of solid NaCl.
4. ***Standard EDTA Solution 0.01M:*** Dissolve 3.723 gm EDTA salt and dilute to 1000 ml.
5. ***Standard $ZnSO_4$ Solution 0.01M:*** Dissolve 1.7945 gm $ZnSO_4$ H_2O (Molecular Weight: 179.45) salt and dilute to 1000 ml.

PROCEDURE

(i) Pipette 25 ml of an aluminium ion solution (approximately 0.01 M) into a conical flask and from a burette add a slight excess of 0.01 M EDTA solution: adjust the pH to between 7 and 8 by the addition of ammonia solution (test drops on phenol red paper or use a pH meter).

(ii) Boil the solution for a few minutes to ensure complete complexation of the aluminium: cool to room temperature and adjust the pH to 7-8.

(iii) Add 50 mg of solochrome black/potassium nitrate mixture and titrate rapidly with standard 0.01 M zinc sulphate solution until the colour changes from blue to wine red.

(iv) After standing for a few minutes the fully titrated solution acquires a reddish-violet colour due to the transformation of the zinc dye complex into the aluminium–solochrome black complex; this change is irreversible, so that over-titrated solutions are lost.

(v) Every ml difference between the volume of 0.01 M EDTA added and the 0.01 M zinc sulphate solution used in the back-titration corresponds to 0.2698 mg of aluminium.

(vi) The standard zinc sulphate solution required is best prepared by dissolving about 1.63 g (accurately weighed) of granulated zinc in dilute sulphuric acid, nearly neutralising with sodium hydroxide solution, and then making up to 250 ml in a graduated flask; alternatively, the requisite quantity of zinc sulphate may be used. In either case, de-ionised water must be used.

OBSERVATIONS AND CALCULATIONS

Al^{+3} Solution Vs $ZnSO_4$ Solution

S. NO.	Volume of Al^{+3} sample (ml) (V_1)	Burette Reading		Volume of $ZnSO_4$ added (ml) (V_2)
		Initial (ml)	Final (ml)	

The amount of Al^{+3} present in the given solution = [Volume of 0.01M EDTA added in each titration − volume of Zn solution consumed for each titration (v_2)] × 0.2698 × $\frac{\text{Total volume of } Al^{+3} \text{ solution}}{V_1}$

RESULT

Amount of aluminium present in the given sample = ——— mg/l

EXPERIMENT - 7 Estimation of Ferrous by Permanganometry

AIM

To estimate the amount of Fe^{+2} iron present in given solution by permanganometry.

APPARATUS

1. Standard flask 2. Pipette 3. Conical flask 4. Burette

CHEMICALS

1. Oxalic acid (N/20) 2. Dil. H_2SO_4 3. $KMnO_4$ (N/20) 4. Iron solution (N/20)

THEORY

In the presence of Dil. H_2SO_4, $KMnO_4$ oxides $FeSO_4$ to $Fe_2(SO_4)_3$ in the cold condition.

$$2\,KMnO_4 + 3H_2SO_4 \longrightarrow K_2SO_4 + 2\,MnSO_4 + 3H_2O + 5\,(O)$$

$$10\,FeSO_4 + 5H_2SO_4 + 5\,(O) \longrightarrow 5Fe_2(SO_4)_3 + 5H_2O$$

Preparations of standard solution of oxalic acid: Weigh the given oxalic acid (0.63049)

$$\text{Eq. wt. of oxalic acid, } 2H_2O = \text{Mol. } \frac{wt}{2} = \frac{126.08}{2} = 63.04$$

$$\text{Eq. wt. of } KMnO_4 = \frac{\text{mol. wt}}{5} = \frac{158.03}{5} = 31.60$$

accurately and transfer in to a flask by using funnel. Dissolve the salt with minimum quantity of water and make up to the mark, shake well.

$$\text{Normality} = \frac{\text{weight} \times 1000}{\text{Eq. wt.} \times \text{Vol in (ml)}}$$

Oxalic acid solution molarity = ________ M

PROCEDURE

Standardization of $KMnO_4$

(i) Pipette out 20 ml of oxalic acid solution into a clean conical flask after rinsing the pipette with same, add 30 ml of dil. H_2SO_4 and heat the solution to gentle boiling titrate this hot solution slowly with $KMnO_4$ with constant stirring.

If $KMnO_4$ added rapidly a brown precipitate of hydrated manganese dioxide is formed. If titrate with $KMnO_4$ to the just permanent faint pink colour.

$$2KMnO_4 + 3H_2SO_4 + 5H_2C_2O_4 \longrightarrow K_2SO_4 + 2MnSO_4 + 8H_2O + 10\,CO_2$$

Standardization Oxallic Acid Vs $KMnO_4$ Solution

Vol of Oxalic acid Solution (V_1) 20 (ml)	Burette reading		Volume of $KMnO_4$ (V_2) (ml)
	Initial (ml)	Final (ml)	

Normality of $H_2C_2O_4$ = N_1 Normality of $KMnO_4$ = N_2

Volume of $H_2C_2O_4$ = V_1 = 20 ml volume of $KMnO_4$ = V_2

$$\text{Normality of } KMnO_4 = \frac{N_1V_1}{V_2} = \text{———} N$$

Estimation of Ferrous

(ii) Make up the given solution up to the mark with distilled water and shake well to get the uniform concentration and pipette out 20 ml of ferrous solution into a clean conical flask and add 20 ml of H_2SO_4, titrate the solution with $KMnO_4$ till the color of the solution changes from colorless to pale pink. Repeat the titrations for concurrent results.

$$MnO_4^- + 8H^+ + 5Fe^{+2} \rightarrow 5Fe^{+3} + Mn^{+2} + 4H_2O$$

$$\text{Eq. wt. of Mohr salt solution} = \frac{\text{mol. wt}}{1} = \frac{392.14}{1} = 392.14$$

Fe^{+2} Solution Vs $KMnO_4$ Solution

S.NO	Volume of unknown 20 (ml)	Burette reading		Volume of $KMnO_4$ (ml)
		Initial	Final	

$$\text{Nomality of given Iron Solution} = \frac{\text{Normality of } KMNO_4 \times \text{volume of } KMNO_4}{20}$$

RESULT

The amount of Iron present in given solution is ——— gm

EXPERIMENT – 8

Preparation of Standard Potassium Dichromate and Estimation of Ferric Iron

AIM

To estimate the amount of ferric iron in given solution.

APPARATUS

1. Pipette
2. Standard flask
3. Burette
4. Conical flask

CHEMICALS

1. 0.12 N Mohr's slat solution
2. 0.05 N $K_2Cr_2O_7$
3. Conc. H_2SO_4
4. Conc. HCl
5. $SnCl_2$
6. Mercuric chloride
7. Diphenyl amine indicator.

THEORY

Ferric iron is reduced to ferrous iron by stannous chloride in presence of hydrochloride and to maintain temperature 70-90°C. The excess stannous chloride is removed by addition of Mercuric chloride. The Fe^{+2} is formed. This ferrous iron is titrated against potassium dichromate. The potassium dichromate is standardized by using Mohr's salt.

$$2\,FeCl_3 + SnCl_2 \rightarrow 2\,FeCl_2 + SnCl_4$$

$$SnCl_2 + 2HgCl_2 \rightarrow Hg_2Cl_2 + SnCl_4$$

$$K_2Cr_2O_7 + 4\,H_2SO_4 \rightarrow K_2SO_4 + Cr_2(SO_4)_3 + 4\,H_2O + 3(O)$$

$$3[2FeSO_4 + H_2SO_4\,(O) \rightarrow Fe_2(SO_4)_3 + H_2O]$$

Mol. wt. of $K_2Cr_2O_7$ = 294.21

$$\text{Eq. wt. of put dichromate } K_2Cr_2O_7 = \frac{294.21}{6} = 49.03$$

Mol. wt. of Mohr salt $FeSO_4\,(NH_4)_2\,SO_4.\,6H_2O$ = 392.14

$$\text{Eq. wt. of Mohr salt} = \frac{392.14}{1} = 392.14$$

PREPARATION OF REAGENTS

1. 0.12N Standard Mohr's salt Solution

Weigh accurately about 4.732 gm of Mohr's salt and transfer into 100 ml standard flask with help of funnel and dissolve salt in concentrated Sulphuric acid and make up to the mark with distilled water.

2. 0.05N potassium dichromate

Weigh accurately about 1.245 gm of potassium dichromate and dissolved in 100 ml with distilled water up to the mark i.e 0.05N potassium dichromate solutions.

PROCEDURE

(i) **Standardization of Potassium Dichromate**

Rinse pipette with Mohr's Salt solution and pipette out 20 ml Mohr's salt solution into a clean conical flask and add indicator Diphenyl amine 2 drops to the solution and titrate with potassium dichromate solution. The end point is color less to blue violet. Repeat the titration to get the concurrent readings and calculate the Molarity of Potassium dichromate.

(ii) **Estimation of Ferric Iron**

Given unknown ferric solution is make up to the mark with distilled water in a 100 ml standard flask and shaking well to get the uniform solution.

(iii) Pipette out 20 ml of the above solution into a clean conical flask add 20 ml of concentrated HCl and heat the solution to boiling, the color of solution changes to pale yellow.

(iv) Add tin Chloride solution drop wise into hot Fe solution till the yellow color disappears, cool the solution under a tap and add 10 ml of solution of mercuric Chloride is one portion, silky white precipitation of $HgCl_2$ is obtained (if the solution turns black due to the formation of finely divided Mercury discord the solution and take fresh solution and reduced)

(v) Add 20 ml of acid mixture and 3-4 drops of Diphenyl Amine indicator, titrate against $K_2Cr_2O_7$ from the burette till the green color changes to blue violet.

(vi) Repeat the titrations to the concurrent readings and calculate the Normality of the Fe^{+3} iron.

OBSERVATIONS AND CALCULATIONS

Standardization of potassium dichromate

Volume of Mohr's salt V_1 = 20 ml   Volume of $K_2Cr_2O_7$ V_2 = ?(from table)

Normality of Mohr's salt N_1 = 0.12N   Normality of $K_2Cr_2O_7$ N_2 =?

$$N_1V_1 = N_2V_2$$

Mohr's salt Solution Vs $K_2Cr_2O_7$ Solution

S. NO.	Volume of Mohr's salt (ml)	Burette readings		Volume of $K_2Cr_2O_7$ (ml)
		Initial (ml)	Final (ml)	
1	20			
2	20			
3	20			

$$\text{Normality of } K_2Cr_2O_7 = N_2 = \frac{N_1V_1}{V_2}$$

Determination of normality of Ferric iron Solution

Volume of Fe^{+3} solution V_3 = 20 ml   Volume of $K_2Cr_2O_7$ V = (from table)

Normality of Fe^{+3} solution N_3 = ———   Normality of $K_2Cr_2O_7$ N_2 =?

$$\text{Normality of unknown } Fe^{+3} \text{ solution } (N_3) = \frac{\text{Normality of } K_2Cr_2O_7 \text{ solution} \times \text{Volume of } K_2Cr_2O_7}{20}$$

Fe^{+2} Solution Vs $K_2Cr_2O_7$ Solution

S. NO.	Volume of Fe^{+3} sample (ml)	Burette Reading		Volume of $K_2Cr_2O_7$ added (ml) (V)
		Initial (ml)	Final (ml)	
1				
2				
3				

$$\text{Amount of ferric iron present in the given solution} = \frac{\text{Normality of } Fe^{+3} \times 55.85}{10} \text{ gm/100 ml}$$

RESULT

Amount of ferric ion present in the given soluton is = ———

EXPERIMENT – 9

Estimation of Zinc by Potassium Ferrocyamide (Precpitation Titration)

AIM

To estimate the amount of zn present in given solution by titrating against standard potassium Ferrocyanide solution.

APPARATUS

1. Standard flask
2. Pipette
3. Conical flask
4. Burette.

CHEMICALS

1. Dil. H_2SO_4
2. Ammonium sulphate
3. Diphenyl benzedene indicator
4. Potassium ferrocyanide
5. Zinc solution.

THEORY

Zn in neutral medium or acidic medium reacts with potassium Ferrocyanide to form a sparingly soluble potassium zinc ferro cyanide.

$$3\,Zn^{2+} + 2\,[Fe(CN)_6]^{4-} \longrightarrow (Zn_3\,[Fe(CN)_6]_2)^{-2}$$

Diphenyl benzedine is used as an indicator.

PROCEDURE

(i) ***Preparation of Standard Zn Solution :*** Weigh the given salt accurately and transfer into 100 ml flask and dissolve in 10 ml dil. H_2SO_4 and make up the solution with distilled water up to the mark and shake it well for uniform concentration. ($ZnSO_4 7H_2O$ mol.wt = 287.55)

(ii) ***Standardisation of $K_4[Fe(CN)_6]$:*** Rinse the pipette with Zn solution and Pipette out 20 ml of same into a conical flask, add 20 ml of same into a conical flask, add 20 ml of distilled water. 20 ml of H_2SO_4 followed by addition Amonium sulphate solultion and 2, 3 drops of diphenyl benzedene indicator. Titrate the solution slowly with vigrous shaking against K_4 [Fe (CN)]$_6$ until colour changes blue violet to pale green.

(iii) ***Estimation of Zinc:*** Make up given solution up to the mark with distilled water and pipette out 20 ml into a clean conical flask. Followed by addition of 20 ml distilled water add 20 ml of H_2SO_4, followed by addition Ammonium sulphate and add 2, 3 drops of indicator. Titrate against $K_4[Fe\,(CN)_6]$ with vigrous shaking till the colour changes blue violet to pale green.

OBSERVATIONS AND CALCULATIONS

$$\text{Molarity of zinc solution} = \frac{\text{Weight}}{\text{M.wt}} \times \frac{1000}{100} = M_1$$

Standardization Zn Solution Vs $K_4[Fe(CN)_6]$ Solution

S.No.	Volume of Std. Zn solution	Burette reading		Volume of $K_4[Fe(CN)_6]$ (ml)
		Initial (ml)	Final (ml)	
1	20ml	0		
2	20 ml	0		
3	20 ml	0		

Molarity of Zn Solution = M_1, Molarity of $K_4[Fe(CN)_6]$ = M_2

Volume of Zn Solution = 20 ml, Volume of $K_4[Fe(CN)_6]$ = V_2

$$\frac{M_1V_1}{n_1} = \frac{M_2V_2}{n_2}$$

$$M_2 = \frac{M_1V_1N_2}{V_2N_1}$$

Unknown Zn Solution Vs $K_4[Fe(CN)_6]$ Solution

S.No.	Volume of unknown ZnSol	Burette reading		Volume of $K_4[Fe(CN)_6]$ (ml)
		Initial (ml)	Final (ml)	
1	20ml	0		
2	20 ml	0		
3	20 ml	0		

Molarity of unknown zinc solution = M_3

Volume of unknown zinc solution = V_3 = 20 ml

Molarity of $K_4[Fe(CN)_6]$ solution = M_4

Volume of $K_4[Fe(CN)_6]$ solution = V_4

$$\frac{M_3V_3}{n_3} = \frac{M_4V_4}{n_4}$$

$$M_3 = \frac{M_4V_4n_3}{n_4V_3}$$

Amount of Zn in given solution = Molarity of unknown Zn solution = $\frac{(M_3) \times \text{Mol. wt of Zn} \times 65.39}{10}$

RESULT

Amount of Zn present in given solution = ——— g.

Experiment – 10 Estimation of Calcium in Limestone by Permanganometry

AIM

To estimate the calcium in limestone by permanganometry.

APPARATUS

1. Conical flask
2. Beaker
3. Glass funnel
4. Burette
5. Watch glass
6. Pipette

CHEMICALS

1. 0.1 N $KMnO_4$ solution
2. 0.1 N Oxalic acid
3. 8% Ammonium oxalate
4. NH_4Cl
5. Ammonia
6. 5 N H_2SO_4
7. Dilute HCl
8. 6N HCl
9. Dil. H_2SO_4

THEORY

Lime stone essentially consists of calcium carbonate but is generally associated with small quantities of magnesium carbonate. Dolomite is an equimolecular compound of calcium carbonate and magnesium carbonate. These ores are generally contaminated with minor amounts of oxides of iron, aluminium, silicon (free or combined) and organic matter. The minerals have large number of industrial applications and the quality of the mineral as well as its suitability for a particular commercial application depends upon the amounts of $CaCO_3$, $MgCO_3$ and the other constituents present. Hence complete analysis is often necessary.

In this method, the lime stone powder is dissolved in hydrochloric acid and the calcium present in the solution is precipitated as calcium oxalate using oxalic acid or ammonium oxalate in presence of ammonia. The precipitated calcium oxalete after washing is treated with dilute sulfuric acid and the oxalic acid liberated is titrated with a standard $KMnO_4$ solution. From the volume of $KMnO_4$ required for the titration, the amount of Ca in the ore can be calculated:

$$1 \text{ ml of } 1N\ KMnO_4 \equiv 0.020 \text{ g of Ca}$$
$$\equiv 0.028 \text{ g of CaO}$$
$$\equiv 0.050 \text{ g of } CaCO_3$$

Reactions Involved

$$CaCO_3 + 2HCl \longrightarrow CaCl_2 + H_2O + CO_2$$

$$CaCl_2 + \underset{\text{ammonium oxalate}}{(NH_4)_2C_2O_4} \longrightarrow \underset{\text{calcium oxalate}}{CaC_2O_4} + 2NH_4Cl$$

$$CaC_2O_4 + H_2SO_4 = CaSO_4 + \underset{\text{oxalic acid}}{H_2C_2O_4}$$

$$2KMnO_4 + 3H_2SO_4 = K_2SO_4 + 2MnSO_4 + 5\,'O' + 3H_2O$$

$$5H_2C_2O_4 + 5'O' = 5H_2O + 10CO_2$$

PREPARATION OF REAGENTS

1. ***$KMnO_4$ solution:*** Prepare approximately N/10 $KMnO_4$ solution by dissolving 3.16 g per litre of water. Standardise this solution with the standard oxalic acid solution.
2. ***Standard oxalic acid (N/10):*** Dissolve 6.3 g of the pure $H_2C_2O_4$ $2H_2O$ in water and make up to 1 litre in a measuring flask. This is used to standardise the $KMnO_4$ solution.
3. *8% Ammonium oxalate solution*
4. NH_4Cl
5. Ammonia
6. ***5 N H_2SO_4:*** Add slowly 139 ml of concentrated H_2SO_4 (36 N. 98%) to some amount of distilled water. After cooling make up the volume to 1 litre with distilled water.
7. **Dilute HCl**
8. ***6 N HCl:*** Dilute 516 ml of concentrated HCl (11.6 N. 36 %) to 1 litre with distilled water.
9. **Dilute H_2SO_4**

PROCEDURE

(i) Weigh accurately 1 gm of the lime stone into a breaker and treat it with a little dilute HCl to dissolve the ore. Take care to see that there is no loss of the solution due to quick evolution of CO_2 by covering the beaker with a watch glass.

(ii) After all the ore is dissolved, wash the watch glass with water into the same beaker. Render the solution alkaline by adding ammonia.

(iii) Add 1 g of NH_4Cl and stir the solution to dissolve it.

(iv) Continue the stirring and add excess of 8% ammonium oxalate solution.

(v) Boil the solution and allow the calcium oxalate precipitate to settle down.

(vi) Decant the supernatant liquid through a filter paper, wash the precipitate several times with hot distilled water containing a little ammonium hydroxide until the filtrate is free from chloride and oxalate ions (by testing with $AgNO_3$ for chloride and hot acidulated dilute $KMnO_4$ for oxalate). Preserve this filtrate for the determination of magnesium, if necessary, as magnesium pyrophosphate.

(vii) Place the beaker containing the precipitate under the funnel and add 20 to 25 ml of 5N H_2SO_4 into the filter paper to dissolve the calcium oxalate present in the filter paper. The acid also dissolves the calcium oxalate precipitate in the beaker. Wash the filter paper with distilled water and collect these washings also into the same beaker.

(viii) Transfer this solution quantitatively into a 250 ml measuring flask, make it up to the mark and shake the solution thoroughly.

(ix) Take 25 ml of the solution into a conical flask, add 10 ml of dil. H_2SO_4, heat the solution to about 70°C and titrate against the standard $KMnO_4$ solution (which in turn was standardised with standard oxalic acid solution first.)

OBERSERVATIONS AND CALCULATIONS

End point is appearance of pink colour.

Wt. of ore taken = 1 gm

Volume of the oxalic acid solution made up = 250 ml

Volume of oxalic taken for each titration = 25 ml

Volume of $KMnO_4$ solution rundown = V_1 ml

∴ Normality of $KMnO_4$ = N_1

We know that

$$1 \text{ ml of } 1N\ KMnO_4 \equiv 0.02 \text{ gm of Ca}$$

$$\therefore \quad V_1 \text{ ml } N_1\ KMnO_4 \equiv 0.02 \times V_1 \times N_1 \text{ gm of Ca}$$

$$= A \text{ gm of Ca}$$

Limestone Solution Vs $KMnO_4$ Solution

S. NO.	Volume of Limestone solution (ml)	Burette Reading		Volume of $KMnO_4$ (ml)
		Initial (ml)	Final (ml)	
1				
2				
3				
4.				

Hence, 25 ml of the solution contains A gm of Ca

∴ 250 ml of the solution contains 10 × A gm of Ca

Now, 1gm of the ore contains = 10 × A, g of Ca = B gm and Ca (where 10 × A = B)

$$\therefore \% \text{ of Ca present in the ore} = B \times 100 \text{ of Ca} \equiv B \times 100 \times \frac{100}{40} \text{ of } CaCO_3.$$

RESULT

Percentage purity of Lime stone is________

EXPERIMENT – 11 Determination of Percentage of Copper in Brass

INTRODUCTION

In volumetric methods employing iodimetry, an excess of iodide ions are added to the oxidizing substance to be determined and the quantity of iodine liberated (equivalent to the amount of the oxidizing agent) is determined by titration with a sodium thiosulfate solution of known strength. The reaction between iodine and thiosulfate ion is given by,

$$2S_2O_3^{2-} + I_2 \longrightarrow S_4O_6^{2-} + 2I^-$$

The most widely used indicator in iodimetry is an aqueous suspension of starch, which imparts an intense blue color to a solution containing a trace of iodine in the presence of iodide ion due to the formation of a complex.

The principle involved in iodimetric titrations is the same as that in oxidation-reduction reactions. Iodide ion is a moderately good reducing agent and the half reaction, $I_2 + 2e^- \longrightarrow 2I^-$, has a potential of + 0.54 volt. All oxidizing agents which have potentials greater than 0.54 volt are capable of oxidizing iodide ions to iodine.

In this experiment, cupric ions (Cu^{++}) will be reduced to cuprous ions (Cu^+) by iodide. The oxidation potentials of the two half cells are:

$$Cu^{+2} + e^- \longrightarrow Cu^+ \quad \ldots E^0 = +0.15 \text{ volt}$$

$$2I^- \longrightarrow I_2 + 2e^- \quad \ldots E^0 = -0.54 \text{ volt}$$

on mixing Cu^{+2} with I^-, one of the reactions taking place will be, $Cu^{+2} + I^- + e^- \longrightarrow CuI\ (s)$. Because of the high positive value of the E^0 for the reaction (0.86 volt), the equilibrium of the reaction, $2Cu^{+2} + 4I^- \rightleftharpoons 2CuI + I_2$ will lie reasonably far to the right.

A brass in an alloy which consists mostly of copper and zinc; lead, iron and tin will be present in small amounts. The method involves dissolution of brass in nitric acid, removal of the nitrate by fuming and sulfuric acid, adjustment of pH by ammonium hydroxide, complexing iron by phosphoric acid and finally, titrating copper.

With nitric acid (in the dissolution process), tin is converted to metastannic acid, $SnO_2.4H_2O$, while lead, zinc and copper are oxidized to the soluble divalent salts; iron is converted to the ferric state. Fuming with sulfuric acid redissolves metastannic acid and precipitates bead as $PbSO_4$, leaving the other elements in solution. Out of these elements, copper and iron are reducible by iodide, but the interference of iron is eliminated by complexing it with phosphate ion.

AIM

To determine the percentage of copper in a sample of brass by iodimetry.

APPARATUS

1. Burette
2. Pipette
3. Beakers
4. Watch glass
5. Water bath
6. Conical flask
7. Volumetric flask

CHEMICALS

1. Ammonia
2. Acetic acid
3. KI
4. Hypo solution ($Na_2S_2O_3$)
5. Starch indicator
6. 6N HNO_3
7. Concentrated H_2SO_4
8. 6N H_2SO_4
9. H_3PO_4
10. KSCN

THEORY

Any Cupric salt in neutral medium when treated with potassium iodide forms a white precipitation of Cuprous iodide and to line is set free quantitatively. The liberated iodine is titrated against Hypo using starch as the indicator.

$$CuSO_4 + 2KI \longrightarrow Cu\,I_2 + K_2SO_4$$

Cuprous iodide being unstable decompose as follows.

$$2\,Cu\,I_2 \longrightarrow Cu_2\,I_2 + I_2$$

$$I_2 + 2\,Na_2S_2O_3 \longrightarrow Na_2S_4O_6 + 2Na\,I$$

Therefore $$2\,Cu\,SO_4 \equiv I_2 \equiv Na_2S_2O_3$$

The equivalent weight of Copper sulphate or copper is its formula weight i.e. 249.7 gm or 63.54 gm respectively.

PREPARATION OF REAGENTS

1. ***6N H_2SO_4:*** Add slowly 166.8 ml of concentrated H_2SO_4 (36 N, 98%) to some amount of distilled water. After cooling make up the volume to 1 litre with distilled water.
2. ***6N HNO_3:*** Dilute 370.8 ml of conc. HNO_3 (16.2 N, 72%) to 1 litre with distilled water.

PROCEDURE

(i) Dry and clean a piece of brass. Cut it into small pieces and weigh about 0.3 g of the sample accurately to the fourth decimal place. Put it in a conical flask and add about 10 ml of 6N HNO_3. Warm the solution under a hood until the brass is completely dissolved.

(ii) Add 10 ml of concentrated H_2SO_4 and evaporate on a sand bath to copious white fumes.

(iii) Allow the mixture to cool and then, add carefully (one ml at a time from the side of the flask) 20 ml of water. Boil for 1 or 2 minutes and cool.

(iv) While vigorously shaking, add liquor ammonia dropwise until the first blue color of cupric ammonia complex appears.

(v) Add 6N H_2SO_4 dropwise until the dark blue color just disappears. Then, add 2 ml of syrupy phosphoric acid.

(vi) Cool it to room temperature and transfer the solution to a 100 ml volumetric flask. Make the volume up to 100 ml.

(vii) Take 25 ml of this solution in a 125 ml conical flask and add about one gram of solid potassium iodide.

(viii) Titrate immediately with standard thiosulfate solution until there is a faint color of iodine.

(ix) Add 3 ml of 1% starch solution and titrate until the blue color begins to fade. Add 0.5 gm of KSCN and complete the titration.

(x) Repeat the titration at least twice by taking 25 ml of the solution each time.

OBSERVATIONS AND CALCULATIONS

Cu^{+2} Solution Vs Thiosulfate Solution

S. No.	Volume of Cu^{++} solution	Burette reading		Volume of thiosulfate solution required
		Initial (ml)	Final (ml)	
1	25 ml			Y_1 ml
2	25 ml			Y_2 ml
3	25 ml			Y_3 ml
		Average or the best titre value		**Y ml**

Weight of the watch glass = W_1 g

Weight of watch glass + brass sample = W_2 g

Weight of the brass sample = $(W_2 - W_1)$g

Strength of sodium thiosulfate solution = pN

Note: There is a tendency for the blue colour to return, so that first complete disappears of the blue colour for about 10 seconds may be taken as end point.

Normality of Cu^{++} in solution = $\frac{p \times Y}{25}$ N = z N

Here, equivalent weight of copper = atomic weight

Amount of copper per liter = z × atomic weight (63.5)

Amount of copper in 100 ml = $z \times \text{atomic weight} \times \frac{100}{1000}$ = b grams

Percentage of copper in brass = $\frac{b \times 100}{(w_2 - w_1)} = c$

OR alternatively

Note: 1 ml 1N $Na_2S_2O_3$ (hypo) = 0.06354 g. Cu

x ml YN $Na_2S_2O_3$ (hypo) = 0.06354 × x × 4 gm of Cu

= A gm of Cu

A gm of Cu is present in $(w_2 - w_1)$ gm of Brass

$$\% \text{ of copper in brass} = \frac{A \times 100}{w_2 - w_1}$$

RESULT

% of Copper in Brass = ____________

EXPERIMENT - 12 Estimation of Manganese Dioxide in Pyrolusite

INTRODUCTION

Manganese dioxide occurs in nature in the form of pyrolusite. For many practical applications, a knowledge about the percentage of MnO_2 is essential. Pyrolusite is not only used as a source of manganese but also as an oxidizing agent in many industrial processes (*e.g.*, Weldon process for manufacturing chlorine). For such purposes, the ore is graded on the basis of its "available oxygen content" rather than on its percentage of manganese, since in many cases, the pyrolusite ore contains less available oxygen than that corresponding to the formula MnO_2.

The percentage of MnO_2 is usually determined by treatment with an excess of an acidified solution of a reducing agent, such as sodium oxalate or arsenic (III) oxide.

$$MnO_2 + H_2C_2O_4 + 2H^+ = Mn^{2+} + 2CO_2 + 2H_2O$$

$$2MnO_2 + 2H_3AsO_3 + 4H^+ = 2Mn^{2+} + 2H_3AsO_4 + 2H_2O$$

The excess of reducing agent is determined by titration with standard permanganate solution.

Note: As (III) oxide is somewhat more preferable in this determination than sodium oxalate, because oxalic acid decomposes very slowly at high temperatures into carbon monoxide and carbon dioxide, the decomposition being catalysed by Mn (II) salts. However, the extent of decomposition under ordinary conditions is very small.

AIM

To estimate the amount of MnO_2 present in the given sample of Pyrolusite

APPARATUS

1. Balance	2. Pipette	3. Burette	4. Conical Flask
5. Glass rod	6. Weighing bottle	7. Funnel	8. Baker.

CHEMICALS

1. Pyrolusite 2. Sodium Oxalate 3. Dil. H_2SO_4 4. $KMnO_4$

THEORY

The MnO_2 present in the Pyrolusite sample is reduced to a known excess of standard sodium oxalate as acid medium, the unreacted sodium oxalate is titrated against a standard solution of $KMnO_4$

$$MnO_2 + H_2SO_4 + H_2C_2O_4 \longrightarrow 2CO_2 + 2H_2O + MnSO_4$$

1m of 1N $KMnO_4$ ≡ 1m of 1N $Na_2C_2O_4$ ≡ 0.04346 gm of MnO_2 ≡ 0.01099 gm Mn

1 ml of 1N $KMnO_4$ ≡ 1ml of N . As_2O_3 ≡ 0.04346 gm of MnO_2

STANDARDISATION

Standardise this solution with the standard N/10 oxalic acid solution as follows:

Take 25 ml of the N/10 oxalic acid solution in a conical flask, add 150 ml of 1M H_2SO_4. Titrate with the $KMnO_4$ solution at room temperature until the faint pink colour appears. Warm the solution to 60°C-70°C and continue the titration to a faint permanent pink colour.

Note: Oxalate solutions attack glass and should not be stored for more than a few days.

PROCEDURE (Sodium oxalate method)

(i) Weigh out accurately about 0.2 g of the finely powdered, dry pyrolusite into a 250 ml conical flask and add 50 ml of standard N/10 oxalic acid.

(ii) Add 50 ml of dilute H_2SO_4 (10%) and place a short funnel over the conical flask.

(iii) Boil the contents of the flask *gently* until no black particles are visible in the flask.

(iv) Allow it to cool about 60-70 °C and titrate the excess oxalate with standard N/10 $KMnO_4$ solution.

(v) Repeat the determination with two other samples of similar weight.

(vi) Calculate the amount of sodium oxalate consumed in the reaction and from this the % MnO_2 in the pyrolusite ore.

OBSERVATIONS AND CALCULATIONS

Wt. of the ore taken = 0.2 g.

Volume of N/10 oxalic acid added in the experiment = 50 ml

Standarization of $KMnO_4$ solution

25 ml of N/10 $KMnO_4$ ≡ 25 ml of N/10 $KMnO_4$ solution

∴ Normality of $KMnO_4$ solution ≡ N/10

Volume of N/10 $KMnO_4$ run down = V ml

∴ Vol. of N/10 $KMnO_4$ consumed = (50 − V) ml

(*i.e.*, (50 − V) ml of N/10 $KMnO_4$ was formed from pyrolusite taken)

We know that

$$1 \text{ ml of } 1\text{N } KMnO_4 \equiv 0.04346 \text{ g } MnO_2$$

$$\therefore \quad (50 - V) \text{ ml of N/10 } KMnO_4 \equiv 0.04346 \times (50 - V) \times \frac{1}{10} \text{ g of } MnO_2$$

$$= \text{A gm of } MnO_2$$

0.2 g of the pyrolusite ore contains A g MnO_2

∴ 100 g of the pyrolusite ore contains

$$= A \times \frac{100}{0.2} \text{ g of } MnO_2$$

$$= 65.190 \text{ g of } MnO_2$$

$$\therefore \ \% \ MnO_2 \text{ in the pyrolusite} = \frac{A \times 100}{0.2}\ \% = B\%$$

86.95 g of MnO_2 contains 16 g of avaialable O_2

$$\therefore \ \text{B g of } MnO_2 \text{ contains} = \frac{B \times 16}{86.95} = C\%$$

∴ % of avaialble oxygen in the pyrolusite ore = C%

Note : 1. The available oxygen is that part of O_2 which is available for oxidation of a reducing agent when the ore is treated with an acid.

$$MnO_2 + 2HCl = MnCl_2 + H_2O + \cdot O \cdot$$

$$2HCl + O = H_2O + Cl_2$$

$Na_2C_2O_4$ Solution Vs $KMnO_4$ Solution

S.No.	Volume of sodium oxalate (ml)	Burette reading		Volume of $KMnO_4$ (ml)
		Initial (ml)	Final (ml)	
1.				
2.				
3.				
4.				

RESULTS

Percentage purity of given pyrolusite = ____________

EXPERIMENT – 13 Estimation of Ferric Iron in Cement by Colorimetrically

AIM

To estimate the amount of Iron (ferric) present in the given sample of cement by colorimetry using ammonium thiocyanate as the reagent.

APPARATUS

1. Standard flasks 2. Burerre 3. Pipette 4. Photo calorimeter 5. Beaker.

CHEMICALS

1. Ferrous ammonium sulphate 2. H_2SO_4 3. Dil. $KMnO_4$ 4. HNO_3
5. 40% potassium thiocyanate 6. Conc. HCl 7. 40% ammonium thiocyanate.

THEORY

The quantitative colorimetry is based on the two principle laws of photometry namely Lambert's Law and Beer's law. They can be briefly stated as follows.

LAMBERT'S LAW

Lambert's law states that when a beam of monochromatic light passes through absorbing medium the rate of decrease in intensity with the thickness of the medium is proportional to the intensity of light (I). Mathematically the law can be written in the logarithmic forms.

$$I = I_o . e^{-k_1 t} \quad \text{(or)} \quad 2.303 \log_{10} \frac{I_o}{I} = k_1 t$$

and

$$D = \text{Optical density} = \log \frac{I_o}{I}$$

$$\frac{I_o}{I} = \text{Opacity}$$

where I_o = the intensity of light
I = intensity of light absorbed
K = Constant
t = thickness of the medium (cm)

BEER'S LAW

Beer's law states that the intensity of emitted light decreases exponentially as the concentration of the light absorbing component increased arithmetically. The mathematical form of the law is written as

$$2.30. \log_{10} \frac{I_o}{I} = K_2 C \text{ (or) } I = I_o . e^{-k_2 C}$$

Here K_2 and C are the proportionality constant and concentration of the chromogen respectively.

The combined form of the two laws generally known as Lambert's Beers law or simply Beer's law. According to Lambert's – Beer's law

$$A = \log \frac{I_o}{I} = \in Ct \qquad \text{where } A = \text{absorbance}$$

$\in$ = Molecular extenstion coefficient

C = Conc. in gram moles / lit

$$\text{Transmission } T = \frac{I}{I_o}$$

or Transmittance

Ammonium thiocyanate yields a blood red colour with ferric iron and the colour produced is stable in nitric acid medium. Its optical density is measured in a photo colorimeter ard the concentration of ferric is found from a standard calibration curve.

PROCEDURE

(a) ***To obtain standard calibration curve:***

(i) 0.7022 g. of AR Ferrous Ammonium Sulphate is dissolved in 100 ml. of water and 5 ml of (1:5) Sulphuric Acid.

(ii) To this add dil. potassium permanganate solution through burette until light pink colour appears. The solution is further diluted to one litre such that 1 ml of solution contains 0.1 mg of ferric iron.

(iii) From the above solution take separately 1 ml. 2 ml. 3 ml. 4 ml and 5 ml. into five 100 ml standard flasks. Add 1 ml of HNO_3 and 5 ml of 40% Potassium Thiocyanate solution to all the above samples to get blood red colour, and the solutions are made up to the mark by adding distilled water.

(iv) The optical densities of all the solutions are determined using photo colorimeter.

(v) A graph is plotted by taking the amount of Ferric Iron on X axis and optical density on Y axis. The curve so obtained is called standard calibration curve. (Fig. 1)

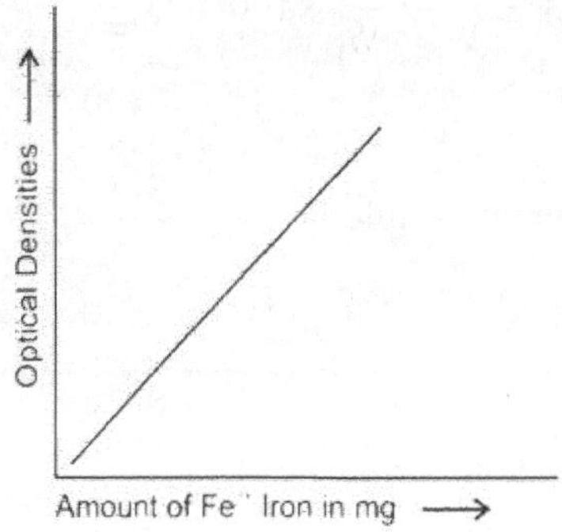

Fig. 1 Model calibration Curve.

The curve is used for determination of Ferric Iron present in the unknown sample.

(b) ***Dissolution of the sample:***

(i) About 0.1 gm of the cement sample is weighed accurately and transferred into a clean and dry 250 ml beaker.

(ii) 5 ml of water is added to moisten the sample. A glass rod is placed and the beaker is covered with a watch glass and about 5 ml of conc. HCl is added drop wise and the solution is heated till the sample dissolves.

(iii) The beaker is kept on small flame and evaporate the solution to almost dryness to expel the excess acid.

(iv) 20 ml of distilled water is added to the beaker to dissolve the contents.

(v) The solution is filtered through No.41 filter paper into a 100 ml standard flask. The funnel is washed twice with 10 ml portions of distilled water into the beaker.

(vi) The funnel is taken out and make up the solution to 100 ml with distilled water. The flask shaken well for uniform concentration.

(c) ***Development of colour:***

(i) 10 ml of the solution prepared above is pipetted out into a 100 ml standard flask, and add to it 1 ml of conc. HNO_3.

(ii) From the burette 5 ml of 40% ammonium thiocyanate is added. Make up the solution to 100 ml with distilled water and shake the flask well for uniform concentration.

(iii) The optical density of the solution is found out by using colorimeter and the concentration of ferric iron from the calibration curve.

OBSERVATION AND CALCULATIONS

Weight of empty bottle = W_1

Weight of weighing bottle + cement = W_2

Weigh of the cement = $W_2 - W_1$

Optical density measured = x corresponding amount of Fe^{++} = Y mg

Percentage of Fe^{+++} in the sample = $\dfrac{Y \times 10}{W_2 - W_1} \times 100$ in mg

RESULT

Percentage of Fe^{3+} present in the cement sample is ———

EXPERIMENT - 14 Estimation of Calcium in Cement

AIM

To estimate the presence of calcium in given cement sample.

APPARATUS

1. Conical flask
2. Burette
3. Beaker
4. Standard flask

CHEMICALS

1. Oxalic Acid N/30
2. Potassium Permanganate N/30
3. Dil. H_2SO_4 (1:5)
4. Concentrated HCl
5. Ammonium Oxalate
6. 10% Ammonium hydroxide

THEORY

The estimation of Calcium is carried out by converting it into Oxalate which is estimate by titrating against standard $KMnO_4$ solution. $KMnO_4$ solution is standardized by using standard oxalic acid solution in acidic medium. $KMnO_4$ oxidizes oxalate ions to CO_2 and gets reduced to Mn (II) ions. Thus the reaction is very slow, the rate is increased on heating 60-70°C.The oxidation is quantitative.

$$Ca^{2+} + 2HCl \rightarrow CaCl_2 + 2H^+$$

$$Ca^{2+} + C_2O_4^{2-} \rightarrow Ca\,C_2O_4 \text{ calcium oxalate}$$

$$2MnO_4 + 5C_2O_4^{2} + 16H^+ \rightarrow 2Mn^{+2} + 10\,CO_2 + 8H_2O$$

PREPARATION OF REAGENTS

Preparation of N/30 Oxalic Acid: 1.05 g of Oxalic Acid is weighed and dissolved in 100 ml distilled water and make up to 500 ml in volumetric flask

Preparation of Potassium Permanganate N/30: 1.05 g of potassium permanganate is dissolved in 100ml distilled water and makes up to 1000 ml in a flask.

Preparation of Ammonium Oxalate saturated solution: 50 g of Ammonium Oxalate is weighed and dissolved in 1000 ml of volumetric flask

1:5 Dil. H_2SO_4: One part of acid + 5 parts of distilled water

10% Ammonium hydroxide: 10 ml of Ammonium hydroxide in 90 ml of distilled water

PROCEDURE

Standardization of $KMnO_4$ solution:

(i) 25 ml of Oxalic acid + 20 ml of Dil. H_2SO_4 are taken in a 250 ml conical flask and heat the contents to 80-100°C and titrate the against $KMnO_4$ solution taken in a burette. Repeat the concurrent values. End point is indicated by permanent pink color.

After Standardization:

(ii) 0.2 g of cement is weighed accurately and transferred into a beaker to this 20 ml of Con. HCl and heat for 30 min over a hot plate. This is digestion process.

(iii) Ca^{+2} is converted into $CaCl_2$ and add 2-3 drops of methyl red indicator. to this 15 ml saturated Ammonium Oxalate solution is added and 10% Ammonium hydroxide solution is added drop wise till the color of the solution is turned to light yellow.

(iv) Calcium is precipitate as in the form of Calcium Oxalate. Precipitate is filter by using whatman No 4 paper. Precipitate is washed with water and dissolved in dil. H_2SO_4 and make up to 250 ml with distilled water.

(v) 25 ml of this solution is taken and titrated with standardized $KMnO_4$ solution by adding dil. H_2SO_4 till a permanent pink color is obtained.

Calculation

wt. of cement taken = 0.2 g

vol. of oxalate solute made up = 250 ml

vol. of oxalate solution taken for each solute = 25 ml

vol. of $KMnO_4$ solution run down = V ml

1 ml of 1N KMnO4 = 0.02 g of Ca

RESULT

Amount of Calcium present in Cement sample is ________

Experiment – 15 Estimation of Calcium Oxide in Cement

AIM

To estimate the amount of calcium oxide present in the given sample of cement solution.

APPARATUS

1. Pipette 2. Conical flask 3. Burette

CHEMICALS

1. EDTA 2. NaOH 3. Diethyl amine 4. Patton and Reeder's indicator 5. Poly vinyl alcohol.

THEORY

Cement contains compounds of calcium, aluminum, magnesium, iron, and insoluble silica. When dissolved in acid, silica remains undissolved. On treating with ammonia, aluminum and iron can be precipitated as their hydroxides and separated. The provided solution of cement, therefore, contains calcium and magnesium ions.

To estimate the calcium content in the given solution, a known volume of it is titrated with standard EDTA solution. In the presence of magnesium, calcium can be titrated with EDTA by using Patton and Reeder s indicator at pH values between 12 and 14. The indicator combines with calcium ions forming a red colored complex.

$$Ca^{2+} + In \rightarrow Ca - In \text{ complex}$$
(Wine red)

Near the end point when free calcium ions are exhausted in the solution, further addition of EDTA dissociates Ca-In complex, consumes the calcium ions. And releases free indicator, which is blue in color. Therefore, the color change is wine red to blue.

The interference of magnesium ions is avoided by precipitating them as hydroxide. This is done by adding 8N NaOH and pH of the solution is maintained at a pH of 12.5 by adding 5 ml of diethyl amine. If the indicator is added before the NaOH solution, satisfactory end point is not obtained because magnesium salts from a lake with the indicator as the pH increases and the magnesium – Indicator Lake is co-precipitated with the magnesium hydroxide. A sharper end point may be obtained by adding 2-3 drops of 1% aqueous poly vinyl alcohol to the sample solution before adding NaOH solution. The poly (vinyl alcohol) reduces the adsorption of the dye on the surface of the precipitate.

The standard EDTA solution is prepared by dissolving a known weight of disodium salt of EDTA in known volume of the solution.

Molecular weight of disodium salt of EDTA = 372.24 gm

PREPARATION OF REAGENT

Standard solution of EDTA: Weigh accurately about 2.5 g of EDTA and transfer into a 250 ml standard flask. Dissolve completely with distilled water (add few drops of NaOH solution if the crystals do not dissolve). Make up the solution to the mark and shake well for uniform concentration.

PROCEDURE

Estimation of calcium oxide in cement solution

(i) Pipette out 25 ml of the given cement solution into a clean conical flask.

(ii) To the flask add 2-3 drops of 1% aqueous poly (vinyl alcohol) and 5 ml of diethyl amine and 5 ml of 8M NaOH solution.

(iii) Allow the solution to stand for 3-5 minutes with occasional stirring.

(iv) Add a test tube of distilled water, a pinch of Patton and Reeder s indicator and titrate against the EDTA solution taken in the burette till the color changes from wine red to blue. Repeat for agreeing values.

OBSERVATION AND CALCULATIONS

Weight of W.bottle + EDTA, W_1 =

Weight of w.bottle alone, W_2 =

Weight of EDTA, $(W_1 - W_2)$ W =

Molarity of EDTA solution M_1 =

Estimation of CaO

In burette : EDTA solution

In conical flask : 25 ml cement solution + 2-3 drops of aq. Poly(vinyl alcohol) + 5 ml of diethyl amine + 5 ml 8N NaOH + 1 t.t distilled water.

Indicator : Patton and Reeder s indicator, a pinch.

Color change : Wine red to blue

Cement Solution Vs EDTA Solution

S.No.	Volume of given cement solution (ml) (V_2)	Burette reading		Volume of EDTA (ml) (V_1)
		Initial (ml)	Final (ml)	
1.				
2.				
3.				

25 ml of cement solution consumes V_1 ml of M_1 M EDTA solution.

$$V_1M_1 = V_2M_2$$

Molarity of the cement solution, $M_2 = \dfrac{V_1M_1}{25}$

Wt of calcium oxide in a litre of the cement solution, W = $M_2 \times$ Mol. Wt of CaO

= $M_2 \times 56.08$

= ________

= ________

Hint: 1 ml of 0.01 m EDTA ≡ 0.4008 mg Ca

≡ 0.5608 mg of Cao

or 1 ml of 1m EDTA ≡ 56.08 mg of CaO.

RESULT

Weight of calcium oxide in a litre of the given sample of cement solution = ——— g.

EXPERIMENT - 16 Estimation of Iron in Cement

AIM

To estimate the amount of iron in cement.

APPARATUS

1. Beaker 2. Pipette 3. Standard flask 4. Burette.

CHEMICALS

1. Conc. HCl 2. Bromine water 3. Stannous chloride 4. Standard Mercuric chloride solution
5. H_3PO_4 6. Diphenyl amine indicator 7. $K_2Cr_2O_7$.

THEORY

Estimation of iron is carried out by titrating it against with $K_2Cr_2O_7$ using Diphenyl Amine as internal indicator. Fe^{+3} salts cannot be oxidized by $K_2Cr_2O_7$. However they can be determined by reduction to Fe^{+2} ions. The reduction can accomplished by using stannous chloride thus,

$$Sn^{+2} + 2\,Fe^{+3} \rightarrow 2\,Fe^{+2} + Sn^{+4}$$

Since there are chances that $SnCl_2$ essential to itself may be oxidized in presence of an oxidizing agent ($K_2Cr_2O_7$) excess of it, it present must therefore be destroyed chemically by adding Mercuric Chloride solution to the reaction mixture which reacts with excess of $SnCl_2$ to form $SnCl_4$ thus,

$$SnCl_2 + 2HgCl_2 \rightarrow SnCl_4 + Hg_2Cl_2$$

$HgCl_2$ as formed above in deposited ion the form of milky white precipitate. The supernant liquid can be titrated with standard $K_2Cr_2O_7$ solution using Diphenyl Amine as indicator.

It is possible that $HgCl_2$ obtained above in the last reaction and may be oxidized by $K_2Cr_2O_7$ solution and can be introduced: error in case of excess $SnCl_2$ is present. However if only a slight excess of $SnCl_2$ is used, the quantity of $HgCl_2$ precipitated is very small and it react with oxidizing agent ($K_2Cr_2O_7$) very slowly, so that no appreciable error is introduced. Therefore $SnCl_2$ should be added drop wise until the yellow color of Ferric salt has disappeared and 2 to 3 more drops of $SnCl_2$ solution is added to ensure complete reduction of ferric to ferrous ion. Use of $K_2Cr_2O_7$ is of particular interest in determination of iron on iron ore. The ore is usually dissolved on HCl and Fe^{+3} are reduced to Fe^{+2} with stannous Chloride solution.

$$Cr_2O_7^{-2} + 6\,Fe^{+3}\ 14\,H^{+} \rightarrow 2\,Cr^{+3} + 6\,Fe^{+3} + 7\,H_2O$$

Diphenyl amine is used as an internal indicator. A blue color is obtained at the end point. The indicator is prepared by dissolving 1gm of Diphenyl amine in 100 ml of con.H_2SO_4. The indicator changes its color to blue violet. When subjected to oxidation by $K_2Cr_2O_7$ it is therefore essentially a reagent for definite range of oxidation (or reduction) potential, just as acid base indicators are reagents for definite pH range in presence of $K_2Cr_2O_7$. The indicator DPA (I) is oxidized first to color less Diphenyl benzidine (II) and is then reversibly oxidized to phenyl benzidine violet. Thus Phenyl amine indicator is suitable whether the potential of the solution is 0.73 Volts to 0.79 Volts at this potential. below this range is reduced form i.e the Diphenyl amine form is more predominant and the solution is therefore colorless. On the other hand at a solution potential of 0.79 Volts, the oxidized form (III) is predominant and the solution appears violet. Between potentials 0.732 to 0.79 V these gradual change in colors of phosphoric acid is desirable for it llower the formal oxidation potential of Fe^{+2} or Fe^{+3} by forming a complex $(Fe\ (HPO_4))^+$.

PROCEDURE

(i) Weigh accurately about 0.2 gm of cement in a 500 ml add 20 ml of con. HCl and heat till the cement is decomposed, add 15 ml of bromine water (convert Fe^{+2} to Fe^{+3}) mixture is heated.

(ii) To remaining solution add stannous chloride ($Fe^{+3} \rightarrow Fe^{+2}$) till the solution is colorless.

(iii) The total volume is made up to 100 ml standard flask, take 10 ml of this solution and add standard mercuric chloride solution.

(iv) A small amount of silky white precipitate Hg_2Cl_2 is formed, to this add H_3PO_4 and Diphenyl amine indicator and titrate against $K_2Cr_2O_7$. Permanent color changes.

(v) Repeat the titration for concurrent readings.

OBSERVATIONS AND CALCULATIONS

Cement Solution Vs $K_2Cr_2O_7$ Solution

S.No.	Volume of Cement solution (ml)	Volume of $K_2Cr_2O_7$ (ml)

Normality of Potassium Dichromate (N_1) = 0.06 N Normality of Fe^{+2} (N_2) =

Volume of Potassium Dichromate (V_1) =? Volume of cement solution (V_2) = 10 ml

$$N_1V_1 = N_2V_2 \qquad N_2 = \frac{N_1V_1}{V_2} = \_\_\_\_\_ N$$

Amount of iron in 0.2 gm of cement = N_2 × Mol.wt of Fe

= _____ × 55.85 gm/lit.

= _______

RESULT

Amount of iron in 0.2 gm of Cement = _________ gm/ lit.

EXPERIMENT – 17

Determination of the Capacity of the Given Cation-exchange Resin

INTRODUCTION

In an ion exchange process, a reversible exchange of ions takes place between the stationary ion exchange phase and the external liquid mobile phase. Zeolites and synthetic organic materials are also used as ion exchangers. Liquid ion exchangers dissolved in water immiscible solvents are also available but they are considered as solvent extraction reagents.

AIM

To deterime the capacity of the given cation-exchange resin (e.g., zeo-karb 225 in the hydrogen form).

APPARATUS

1. Beaker
2. Glass rod
3. Small columns
4. Conical flask
5. Burette

CHEMICALS

1. 2M HCl
2. 0.25 gm cation exchange resin
3. 0.1N NaOH
4. 0.25M sodium thio sulphate
5. Phenolphthalein indicator.

THEORY

A cation exchange resin may be represented by (res.A^-) B^+ ,where res is the basic polymer of the resin. A- is the anion attached to the polymeric frame work . B^+ is the active or mobile cation. Thus a sulphonated poly styrene resin in the hydrogen form would be written as (res. SO_3^- H^+).

Cation exchange resins contain free cations which can be exchanged for cations in solutions.

$$(res.A^-)\,B^+ + C^+\,(SOLUTION) \leftrightarrow (res.A^-)\,C^+ + B^+\,(solution)\,.$$

The experimental conditions are such that the equilibrium is completely displaced from left to right. the ion C^+ is completely fixed on the cation exchanger. If the solution contains several cations (C^+, D^+, and E^+) the exchanger may show different affinities for them . thus making separations possible.

By passing a solution of a neutral salt through the hydrogen form of a sulphonic resin, an equivalency quantity of the corresponding acid is produced by the following reaction

$$2(res.SO_3^-)H^+ + Na_2SO_4(sol) \leftrightarrow 2\,(res.SO_3^-)\,Na^+ + H_2SO_4(sol)$$

The total ion exchange capacity of a resin is dependent upon the total number of ion active groups per unit weight of material, and the grater the number of ions, the grater will be the capacity is usually expressed as milli equivalents per gram of exchange. The break through capacity is often much less than the total capacity of the resin.

The exchange capacity of a cation exchange resin may be measured in the laboratory by determining the number of milli grams equivalents of sodium ion which are absorbed by one gram of the resin in the hydrogen form.

$$2R - H^+ + Na_2SO_4 \rightarrow 2R - Na^+ + H_2SO_4$$

PROCEDURE

(i) Weigh out accurately about 0.25 gm of the given cation exchange resin in a 50 ml beaker.

(ii) Soak with 10 ml of 2M HCl for about 30 min with occasional stirring.

(iii) Decant the acid along with the fine particles of resin if any. and wash the resin with several 10 ml portions of deionized water. until the washings are free from acid.

(iv) Partly fill small column. 15 cm × 1cmwith distilled water. taking care to displace any trapped in from beneath the sintered glass disc. Transfer the resin completely along with portions of distilled water into the column.

(v) Fill a 250 ml separate funnel with 150 ml of 0.25M sodium sulphate. Allow this solution to drip into the column at a rate of about 3 ml per min and collect the effluent in 250 ml conical flask. When all the solution has passed through the column. titrate the effluent with standard 0.1N NaOH using Phenolpthalein as indicator.

CALCULATIONS

The capacity of the resin in milli equivalents per gram is given by

$$av/w$$

where

a = normality of NaOH solutions,

v = volume in ml

w = weight of the resin

RESULT

The capacity of the given resin = ______

EXPERIMENT – 18 Estimation of Copper by Colorimetric Method

AIM

To estimate the copper present in the given sample by colorimetry.

APPARATUS

1. Spectrophotometer 2. Conical flask 3. Pipette

CHEMICALS

1. Bicyclohexanone oxalyldihydrazone solution 2. Ethanol 3. Perchloric acid
4. Conc. HNO_3 5. Conc. Ammonia

THEORY

Small quantities of copper may be determined by the diethyldithiocarbamate method (Section 6.10) or by the 'neo-cuproin' method (Section 6.11), an extraction being necessary in both cases. In another somewhat simpler procedure, the copper is complexed with biscyclohexanone oxalyldihydrazone and the resulting blue colour is measured by a suitable spectrophotometer within the range 570–600 nm (orange filter). The solution measured should contain not more than 100 µg of copper.

REAGENTS

Bicyclohexanone oxalyldihydrazone solution (copper reagent): Dissolve 0.1 g of the solid reagent in 10 ml ethanol (or industrial methylated spirit) and 10 ml hot water, and dilute to 200 ml. Filter, if necessary.

Synthetic standard solution (for analysis of steel): Dissolve an appropriate weight of pure iron (Johnson Matthey) in a mixture of equal volumes of concentrated hydrochloric acid and concentrated nitric acid; with this solution as base, add a suitable amount of copper nitrate solution containing 0.01 g copper per litre.

PROCEDURE (COPPER IN STEEL)

(i) Weigh out accurately a 0.1 g sample of the steel into a 150 ml conical beaker, add 5 ml concentrated hydrochloric acid and 5 ml concentrated nitric acid, and warm gently.

(ii) In the presence of interfering amounts of chromium, add 5 ml perchloric acid, specific gravity 1.70, and evaporate until strong fuming occurs.

(iii) When the sample has dissolved or after the fuming with perchloric acid, cool, add 50 ml cold distilled water, followed by 10 ml acid solution (1:1 HCl/HNO_3).

(iv) Carefully add 10 ml concentrated ammonia solution, sp. gr. 0.88, cool to room temperature, and dilute to 100 ml in a graduated flask.

(v) Return the solution to the original beaker and transfer a 10 ml aliquot to a 100 ml graduated flask. Add 20 ml of the copper reagent, dilute to 100 mL with distilled water, and transfer to a 100 ml dry beaker.

(vi) Allow to stand for 10-15 minutes, and then measure the absorbance with a spectrophotometer.

(vii) Construct a calibration curve using the synthetic standard solution: add the standard copper solution immediately before the reagent.

(viii) Plot the graph between absorbance and concentration, gives a straight line from which concentration of the given solution can be estimated.

OBSERVATIONS AND CALCULATIONS

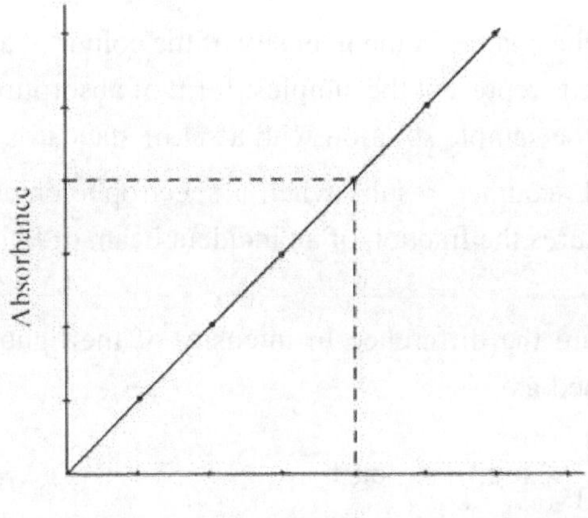

S.No.	Concentration of the solution	Absorbance

RESULT

The concentration of copper solution = ——— moles/litre

EXPERIMENT – 19 Spectrophotometric Determination of Iron

INTRODUCTION

Colorimetric analysis is based on the change in the intensity of the colour of a solution with variations in concentration. Colorimetric methods represent the simplest form of absorption analysis. The human eye is used to compare the colour of the sample solution with a set of standards until a match is found.

An increase in sensitivity and accuracy results when a spectrophotometer is used to measure the colour intensity. Basically, it measures the fraction of an incident beam of light which is transmitted by a sample at a particular wavelength.

There are two ways to measure the difference in intensity of the light beam. One is the percent transmittance, %T, which is defined as:

$$\%T = \frac{I_o}{I} = \log\frac{1}{T} = -\log T$$

For any given compound, the amount of light absorbed depends upon (a) the concentration, (b) the path length, (c) the wavelength and (d) the solvent.

Absorbance is related to the concentration according to the Beer-Lambert law:

$$A = \varepsilon bc$$

where ε is the extinction coefficient ($M^{-1}\ cm^{-1}$), b is the solution path length (cm) and c is the concentration (moles $litre^{-1}$).

Not all substances obey the linear Beer-Lambert law over all concentration ranges. Therefore you will construct a calibration curve that will provide the relationship between concentration and absorbance under the conditions used for the analysis.

AIM

To determine the iron content of an unknown sample by spectrophotometer.

APPARATUS

1. Beakers 2. Pipette 3. Volumetric flasks 4. Spectrophotometer

CHEMICALS

1. O-Phenanthroline solution 2. Hydroxylamine HCl 3. Standard iron solution.

THEORY

Fe^{+2} is reacted with o-phenanthroline to form a coloured complex ion. The intensity of the coloured species is measured using a Spectronic 301 spectrophotometer. A calibration curve (absorbance versus concentration) is constructed for Fe^{+2} and the concentration of the unknown iron sample is determined.

In this experiment you will analyze for iron by reacting Fe^{+2} with o-phenanthroline to form an orange-red complex ion according to the following equation:

Because we are starting with an Fe^{3+} solution and in order to be quantitative, all of the iron must be reduced from Fe^{3+} to Fe^{2+} by the use of an excess of hydroxylamine hydrochloride.

$4\ Fe^{3+}$	$+ 2\ NH_2OH{\bullet}HCl$	$\rightarrow$	$4\ Fe^{2+}$	+	N_2O	$+ 4\ H^+$	$+\ H_2O$
Ferric Iron	Hydroxylamine Hydrochloride		Ferrous Iron		Nitrous Oxide	Proton	Water

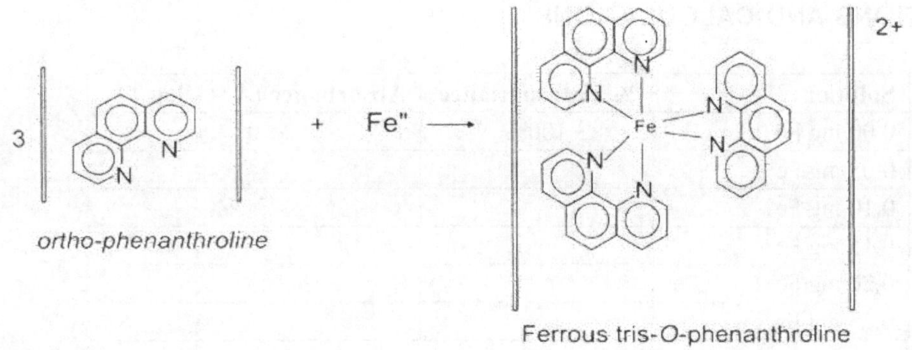

PROCEDURE

(i) Obtain the following quantity of chemicals from the back bench using a labeled beakers.

Unknown iron solution	~15 ml
Standard iron solution (0.250 g/l)	~30 ml
0.3% weight/volume o-phenanthroline solution	~40 ml
10% weight/volume hydroxylamine hydrochloride	~40 ml

(ii) Prepare the following iron calibration solutions by pipetting the indicated amounts of the above iron solution (step 1) into labeled 50 ml volumetric flasks. The first flask is a blank containing no iron.

Concentration of Fe	**Volume to pipette**
0.00 mg Fe	0.00 ml
0.05 mg Fe	4.00 ml
0.10 mgFe	8.00 ml
0.15 mgFe	12.00 ml
0.20 mg Fe	16.00 ml
0.25 mg Fe	20.00 ml

(iii) Pipette 10.0 mL of an unknown sample solution (record the unknown's number) into a 250 ml volumetric flask and dilute to the mark with distilled water. Invert and shake the flask several times to mix the solution.

(iv) Pipette two 25.0 ml aliquots of this solution into two 50 ml volumetric flasks labeled unknown.

(v) Using a 10 ml graduated cylinder, add 4.0 ml of 10% hydroxylamine hydrochloride solution and 4.0 ml of 0.3% o-phenanthroline solution to each volumetric flask.

(vi) Swirl and allow the mixture to stand for 10 minutes.

(vii) Dilute each flask to the mark with distilled water and mix well by inverting and shaking the capped volumetric flasks several times.

(viii) Using the spectrophotometer, carefully measure the percent transmittance of the various solutions in the 50 ml volumetric flasks, including the two unknown solutions. Record your results in the following table.

OBSERVATIONS AND CALCULATIONS

Solution	% Transmittance	Absorbance (A = – log T)
0.00 mg Fe (blank)	100%	0
0.05 mg Fe		
0.10 mg Fe		
0.15 mg Fe		
0.20 mg Fe		
0.25 mg Fe		

1. Prepare a plot of absorbance versus concentration of the known solutions (express the concentration in mg Fe per 50 ml of solution). Draw the best fitting straight line through the points – this is called the Beer-Lambert Law plot.
2. Place the best Absorbance value of each unknown solution onto this plot and determine their concentrations.
3. Calculate the amount of iron in the unknown sample. Express this as mg of Fe per litre of the original unknown solution (mg/l Fe).

   E.g. From the graph you obtain a concentration of 0.10 mg Fe/50 ml

   Since in step 3 we diluted the original sample 25 times and in step 4, 2 more times the concentration of the original sample is therefore:

$$0.10 \frac{\text{mg Fe}}{\text{50 ml}} \times 50\,(\text{dilution factor}) \times \frac{\text{100 mg Fe}}{l} = \frac{\text{100 mg Fe}}{l}$$

   and calculate the relative error using the given formula

$$\text{Relative error} = \frac{\text{Experimental value} - \text{accepted value}}{\text{accepted value}} \times 100\%$$

RESULT

The amount of iron present in the unknown sample = ——— mg/l

EXPERIMENT - 20 Determination of Sodium and Potassium by Flame Photometer

INTRODUCTION

The principle involved in flame photometry or flame emission spectroscopy is that when a solution containing a metallic compound is aspirated into a flame *(e.g.* acetylene burning in air) a vapour containing the metal atoms will be formed. Some of these metal atoms in gaseous state, may be raised to and energy level, which is sufficiently high to permit the emission of radiation, which is characteristic to the metal under investigation. Flame photometers are generally used for the analysis of sodium, potassium, lithium and calcium, because these elements have an easily excited flame spectrum of sufficient intensity for detection by a photocell.

The layout of a simple flame photometer is shown Fig. 1.

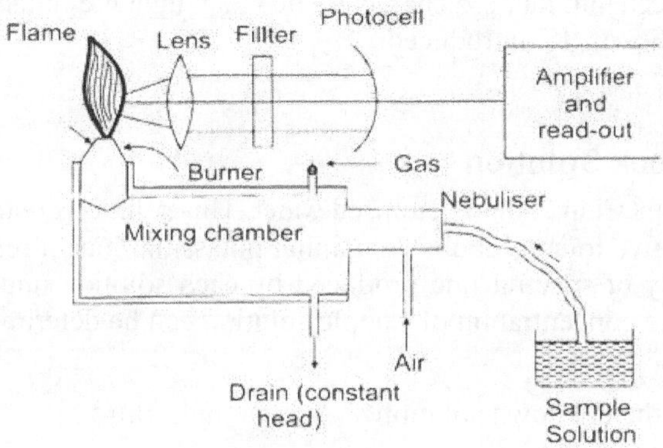

Fig. 1. Lay-out of a simple flame-photometer

AIM

To determine Ca, Na, K and Li by flame photometer.

APPARATUS

1. Flame photometer 2. Standard flask

CHEMICALS

1. KCl 2. NaCl

THEORY

Flame is a much lower energy source than electric means of excitation. Thus flame produces a simple emi spectrum with fewer lines further more relatively cool flames.

ex: Air propane are normally used in radiation is isolated by an optical filter and then converted to an electric signal by photo detector a photomultiplies.

Basic Components

Air at a given pressure into an atomizer and suction this draws a solution of sample into atomizer where it joins air stream as a fine mist and passes into burner at a given pressure and mixture is burnt. Radiation from resulting flame passes through a lens and finally there is an optical filter which permits only the radiation characteristic of elements under investigation to pass through the photodetector.

The output is measure on a digital read out system. The precision of the technique can be improved by using a double beam flame photometer. The internal standard solution based on 'Li' salt is continuously monitored to ensure within precision.

The internal response optics use a 'Li' interference filter. The radio light intensities of either Na/K/Ca can be obtained from appropriate photo detector. The electronic configuration circuit is designed to give a direct reading of 'Na' to concentrations. Besides these facilities it incorporates an integral dilutes which automatically dilutes all samples types. These eliminating time consuming manual pre-dilution process.

DETERMINATION OF POTASSIUM

AIM: To estimate amount of 'K' present in given solution by flame photometry.

THEORY

When 'K: salt (KCl) is introduced into flame, a charecteristic violet light is emitted by flame. The intensity of this light is related to quantity of 'K' introduced.

PROCEDURE

(i) Preparation of Stock Solution

Dissolve 0.4775 gm of KCl in 250 ml deionised water. This solution contains equivalent of dilute the stock solution to give four solutions containing potassium ions. Prepare a calibration curve by measuring intensity of spectral line produced by each solution under identical conditions. With help of this curve concentration of sample solution can be determined.

Result

The concentration of the unknown solution = ——— moles/litre

DETERMINATION OF SODIUM

AIM: To determine amount of sodium present in given solution by flame photometry.

THEORY

When sodium salt is introduced in flame, a characteristic yellow light is emitted by flame. If this light is passed through a dispersing device, lines of characteristics and wave length are observed this intensity of this length is related to quantity of sodium introduced into flame.

PROCEDURE

(ii) Preparation of Stock Solution

Dissolve 0.6868 gm of NaCl in 250 ml deionised water in a graduated flask. This solution contains equivalent of 1000 mg of sodium per litre i.e., 1000 ppm.

Sample preparation: 1 gm of sample in 25 ml of solution

RESULT

The concentration of the unknown solution = ——— moles/litre

EXPERIMENT - 21 Determination of Sulphate by Turbidometry

INTRODUCTION

Suspension of particles in water interfering with passage of light is called turbidity. Turbidity is caused by a wide variety of suspended matter, ranging in size from colloidal to coarse dispersions depending upon the degree of turbulence, and also from pure inorganic substances to those that are highly organic in nature. Turbid water is undesirable from aesthetic point of view in drinking water supplies and may also affect products in industries. The clarity of natural body of water is a measure determinant of the condition and productivity of that system. Turbid water also poses a number of problems in water treatment plants. Consequently turbidity is a single parameter often used for its performance evaluation.

Nephelometric method

Turbidity can be measured by its effect on the scattering of light, which is termed as Nephelometry. Turbidimeter can be used for sample with moderate turbidity and nephelometer for samples with low turbidity. Higher the intensity of scattered lights higher the turbidity.

Turbidity is an expression of the optical property that causes light to be scattered and absorbed rather than transmitted in straight lines through the sample. The standard method for the determination of turbidity has been based on the Jackson candle turbidity meter. However, the lowest turbidity value that can be measured directly on this instrument is 25 units. An indirect method is necessary to estimate the turbidity in the range of 0-5 units: the turbidities of treated water generally fall in this range. Most commercial turbidimeters available for measuring low turbidities give comparatively good indications of the intensity of light scattered in one particular direction, predominantly at right angle to the incident light. These nephelometers are relatively unaffected by small changes in design parameters and are therefore specified as the standard instrument for measurement of low turbidities. Results from nephelometric measurements are expressed as nepholometric turbidity units (NTU).

SULPHATES

AIM

To estimate the sulphate ions in a given sample.

APPARATUS

1. Turbidometer/Nephelometer 2. Sample tube.

CHEMICALS

1. Standard sulphate solution: Dissolve 0.1479 grams of anhydrous Na_2SO_4 sodium sulphate and dilutes to 1000 ml.
2. Barium chloride ($BaCl_2$)
3. Conditioning reagent: Mix 25 ml glycerol with a solution containing 15 ml of concentrated HCl and 150 ml distilled water, 50 ml of 95% Isopropyl alcohol and 37.5 ml of Nacl.

THEORY

Sulphate ions are precipitated as barium sulphate in acidic media (H2SO4 + BaCl2). The absorption of light by this precipitated suspension is measured by spectrophotometer at 420 nm or scattering of light by the neplometers.

Interference: Colour, turbidity and silica in the concentration of 500 ppm interferes in this estimation.

- Filteration is adapted to remove colour and turbidity in presence of organic meter precipitation of barium-sulphate may not be satisfactory.

PROCEDURE

For Water sample

(i) 50 ml of standard sulphate solution was taken and 2.5 ml of conditioning reagent was added and mixed well.

(ii) Flask was stirred and a pinch of Barium chloride crystals was added and continuously stirred for 1 minute and the turbidity was measured by using spectrophotometer at 420 nm.

(iii) 50 ml of the given water and waste water sample was taken and proceeded in the same way as the standard.

(iv) A caliberation chart was prepared for standards by plotting concentration on X-axis and absorbance on Y-axis.

(v) Read the unknown concentration of sulphate in the given water and waste samples on the chart.

For Soil sample

(i) 10 grams of air dried soil or sediment sample was weighed to this 100 ml of distilled. Water was added to make a suspension of 1: 10

(ii) It was filtered through a filter paper [watchman No. 44] now find out the sulphate concentration by using the method described for water.

(iii) From a fraction of same soil or sediment sample the moisture was identified by the method given easilier.

CALCULATIONS

∴ concentration of water sample = concentration × dilution factor

∴ concentration of soil sample = concentration of Sulphate mg/gram

$$\frac{A}{10} \times \frac{B}{X} \times \frac{1}{100 - M}$$

where

A = sulphate estimated in filterate mg/litre

B = Total volume of suspension

x = weight of soil or sediment in grams

M = Moisture content

∴ concentration of soil sample = concentration × dilution factor

RESULT

∴ concentration of SO_4 in water sample = ——— mg/litre

∴ concentration of SO_4 in soil sample = ——— mg/gram

EXPERIMENT – 22 Simultaneous Spectrophotometric Determination of Chromium & Manganese

INTRODUCTION

This section is concerned with the simultaneous spectrophotometric determination of two solutes in a solution. The absorbances are additive, provided there is no reaction between the two solutes. We may write:

$$A_{\lambda 1} = {}_{\lambda_1}A_1 + {}_{\lambda_1}A_2 \qquad \text{.....(1)}$$

$$A_{\lambda 2} = {}_{\lambda_2}A_1 + {}_{\lambda_2}A_2 \qquad \text{.....(2)}$$

where $A_{\lambda 1}$ and $A_{\lambda 2}$ are the measured absorbances at the two wavelengths λ_1 and λ_2; the subscripts 1 and 2 refer to the two different substances, and the subscripts λ_1 and λ_2 refer to the different wavelengths. The wavelengths are selected to coincide with the absorption maxima of the two solutes: the absorption spectra of the two solutes should not overlap appreciably (compare Fig. 1), so that substance 1 absorbs strongly at wavelength λ_1 and weakly at wavelength λ_2, and substance 2 absorbs strongly at λ_2 and weakly at λ_1. Now $A = \varepsilon c l$, where ε is the molar absorption coefficient at any particular wavelength, c is the concentration expressed in mol L^{-1}, and l is the thickness (length) of the absorbing solution expressed in cm. If l is 1 cm:

$$A_{\lambda 1} = {}_{\lambda_1}\varepsilon_1 \,.\, c_1 + {}_{\lambda 1}\varepsilon_2 \,.\, c_2 \qquad \text{.....(3)}$$

$$A_{\lambda 2} = {}_{\lambda_2}\varepsilon_1 \,.\, c_1 + {}_{\lambda 2}\varepsilon_2 \,.\, c_2 \qquad \text{.....(4)}$$

Solution of these simultaneous equations gives:

$$c_1 = \frac{{}_{\lambda_2}\varepsilon_2 . A_{\lambda 1} - {}_{\lambda_1}\varepsilon_2 . A_{\lambda 2}}{{}_{\lambda_1}\varepsilon_1 . {}_{\lambda_2}\varepsilon_2 - {}_{\lambda_1}\varepsilon_2 . {}_{\lambda_2}\varepsilon_1} \qquad \text{.....(5)}$$

$$c_2 = \frac{{}_{\lambda_1}\varepsilon_1 . A_{\lambda 2} - {}_{\lambda_1}\varepsilon_1 . A_{\lambda 1}}{{}_{\lambda_1}\varepsilon_1 . {}_{\lambda_2}\varepsilon_2 - {}_{\lambda_1}\varepsilon_2 . {}_{\lambda_2}\varepsilon_1} \qquad \text{.....(6)}$$

The values of the molar absorption coefficients ε and ε_2 can be deduced from measurements of the absorbances of pure solutions of substances 1 and 2. By measuring the absorbance of the mixture at wavelengths λ_1 and λ_2, the concentrations of the two components can be calculated.

The above considerations will be illustrated by the simultaneous determination of manganese and chromium in steel and other ferro-alloys. The absorption spectra of 0.001 M permanganate and dichromate ions in 1 M sulphuric acid, determined with a spectrophotometer and against 1 M sulphuric acid in the reference cell, are shown in Fig. 1. For permanganate, the absorption maximum is at 545 nm, and a small correction must be applied for dichromate absorption. Similarly the peak dichromate absorption is at 440 nm, at which permanganate only absorbs weakly. Absorbances for these two ions, individually and in mixtures, obey Beer's Law provided the concentration of sulphuric acid is at least 0.5M. Iron(III), nickel, cobalt, and vanadium absorb at 425 nm and 545 nm, and should be absent or corrections must be made.

AIM

To determine the chromium and manganese in an alloy steel by using spectrophotometer.

APPARATUS

1. Kjeldahl flask 2. Standard flask 3. Pipette 4. Conical flask 5. Spectrophotometer.

CHEMICALS

1. Potassium dichromate. 0.002 M. 0.001 M. and 0.0005 M in 1 M sulphuric acid and 0.7M phosphoric (V) acid. prepared from the analytical grade reagents.
2. *Potassium permanganate:* 0.002M. 0.001 M. and 0.0005M in 1 M sulphuric acid and 0.7M phosphoric(V) acid. prepared from the analytical grade reagents.

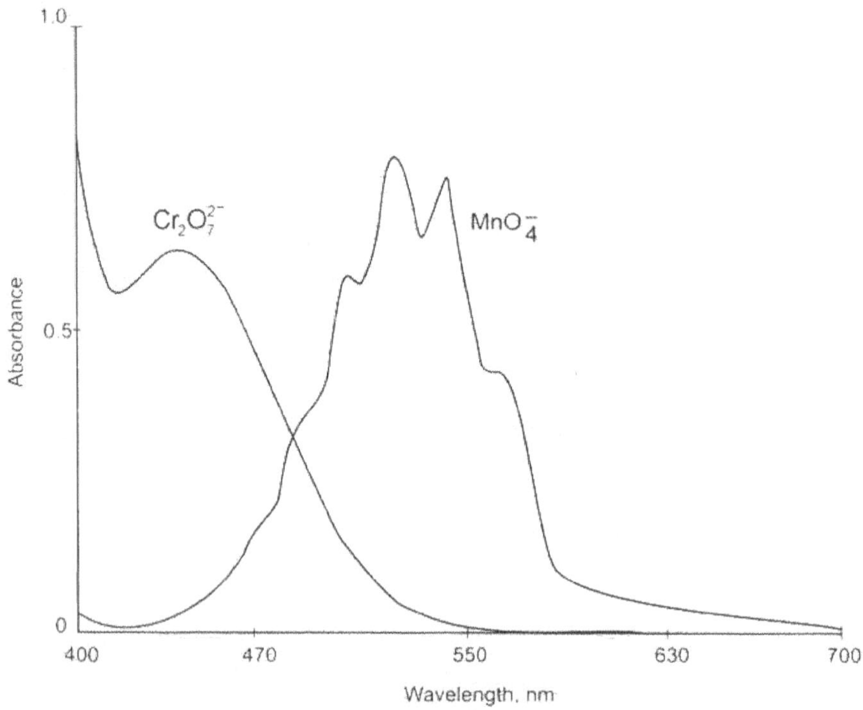

Fig. 1

PROCEDURE

(a) ***Determination of molar absorption coefficients and verification of additivity of absorbances:*** The molar absorption coefficients must be determined for the particular set of cells and the spectrophotometer employed. For the present purpose we may write:

$$A = \varepsilon c l$$

where ε is the molar absorption coefficient. c is the concentration (mol L^{-1}). and l is the cell thickness or length (cm).

(i) Measure the absorbance A of the above three solutions of potassium dichromate and of potassium permanganate, each solution separately, at both 440 nm and 545 nm in 1 cm cells. Calculate ε in each case and record the mean values for $Cr_2O_7^{2-}$ and MnO_4^- at the two wavelengths.

(ii) Mix 0.001 M potassium dichromate and 0.0005M potassium permanganate in the following amounts (plus 1.0 ml of concentrated sulphuric acid), and prepare a set of results similar to those in Table 1, which is a set of typical results included for guidance only.

(iii) Measure the absorbance of each of the mixtures at 440 nm. Calculate the absorbance of the mixtures from:

$$A_{440} = 440\varepsilon_{cr}.c_{cr} + 440\varepsilon_{Mn}.c_{Mn}$$

Table 1 Test of additivity principle with $Cr_2O_7^{2-}$ and MnO_4^- mixture at 440 n

$K_2Cr_2O_7$ solution (ml)	$KMnO_4$ Solution (ml)	A Observed	A Calculated
50	0	0.371	-
45	5	0.338	0.340
40	10	0.307	0.308
35	15	0.277	0.277
25	25	0.211	0.214
15	35	0.147	0.151
5	45	0.086	0.088
0	50	0.057	-

(b) ***Determination of chromium and manganese in an alloy steel:***

(i) Weigh out accurately about 1.0 g of the alloy steel into a 300 mL Kjeldahl flask, add 30 ml of water and 10 mL of concentrated sulphuric acid [also 10 mL of 85 per cent phosphoric (V) acid if tungsten is present].

(ii) Boil gently until decomposition is complete or the reaction subsides.

(iii) Then add 5 ml of concentrated nitric acid in several small portions. If much carbonaceous residue persists, add a further 5 ml of concentrated nitric acid, and boil down to copious fumes of sulphuric acid.

(iv) Dilute to about 100 ml and boil until all salts have dissolved. Cool, transfer to a 250 ml graduated flask, and dilute to the mark.

(v) Pipette a 25 ml or 50 mL aliquot of the clear sample solution into a 250 ml conical flask, add 5 mL concentrated sulphuric acid, 5 mL 85 per cent phosphoric(V) acid, and 1-2 mL of 0.1 M silver nitrate solution, and dilute to about 80 ml.

(vi) Add 5 g potassium persuiphate, swirl the contents of the flask until most of the salt has dissolved, and heat to boiling. Keep at the boiling point for 5–7 minutes.

(vii) Cool slightly, and add 0.5 g pure potassium periodate. Again heat to boiling and maintain at the boiling point for about 5 minutes.

(viii) Cool, transfer to a 100 ml graduated flask, and measure the absorbances at 440 nm and 545 nm in 1 cm cells.

(ix) Calculate the percentage of chromium and manganese in the sample. Use equations (5) and (6) and values of the molar absorption coefficients ε determined above: these will give concentrations expressed in mol L^{-1}, from which values the percentages can readily be calculated. The values listed are the equivalent percentages of the respective constituent to be subtracted from the apparent Cr and Mn percentages for each 1 per cent of the element in question. It can be shown that utilising the known (or determined) molar absorption coefficients ($_{545}\varepsilon_{Cr}$ 0.011; $_{440}\varepsilon_{Cr}$ 0.369; $_{440}\varepsilon_{Mn}$ 0.095):

$$\text{Mn, per cent} = \frac{0.00549V}{W}(0.426\,A_{545} - 0.013\,A_{440})$$

$$\text{Cr, per cent} = \frac{0.01040V}{W}(2.71\,A_{440} - 0.110\,A_{545})$$

for a sample of W grams in a volume of V ml.

RESULT

The concentration of the chromium is found to be = ——— ppm

And, the concentration of the manganese is found to be = ——— ppm.

EXPERIMENT – 23 Inversion of Sucrose by DIGITAL Polarimeter

INTRODUCTION

According to the electromagnetic theory of light, light consists of vibrations in ordinary light, these vibrations occur in all planes perpendicular to its direction of propagation.

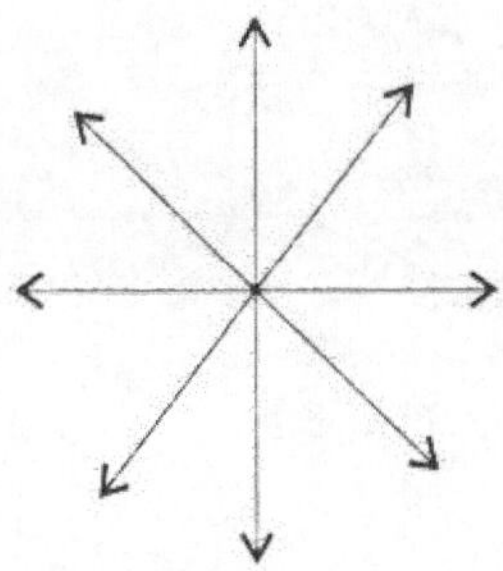

Vibrations of ordinary rays: direction of propagation perpendicular to the plane of paper.

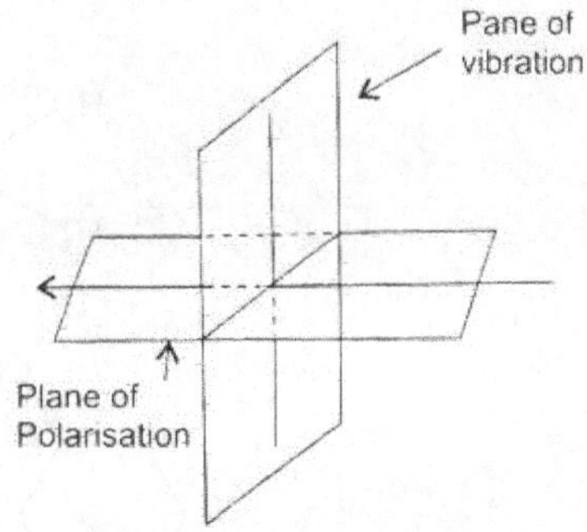

Plane polarised light.

When ordinary light is passed through a Nicol's prism, its vibrations in all directions except the direction of axis of the prism, are cut off. The emergent rays come out having their vibrations in only one plane. *A beam of light whose vibrations occur in only one plane is said to be* **plane polarised light**. The plane in which these vibrations occur is known as the plane of vibrations of the polarised ray, while the plane perpendicular to it is called the *plane of polarisation of the polariser*.

Polarimeters are intstruments which measure the rotation of the plane of polarisaton in certain liquids. The classical instrument is optical device. It has many short commings. It needs a dark room and sodium light. It strains the eyes. It has personal reading error. All these short commings are removed in **DIGITAL POLARIMETER**. It can detect levo and dextro substance directly. It has digital readout for degrees.

Factors controlling values of angles of rotation

The angle through which a substance in the pure or dissolved state, rotates the plane of polarisation of light depends on:

1. The nature of the substance in respect of nature and number of asymmetry centres in its molecule.
2. The nature of solvent.
3. Concentration of solute.
4. The temperature of measurement.
5. The length of optical path in the substance or solutions.
6. The wavelength of the light used.

Causes of optical rotation

According to Pasteur, who discovered the phenomenon of optical rotation, the optical rotation property of a substance is caused by the asymmetry in its molecules. Such substances would have crystals which are mirror images but cannot be superimposed. This situation arises when:

1. the molecule has an asymmetric carbon atom (van't Hoff and Le Bel explanation) Such a carbon, atom has attached to itself 4 atoms or groups of atoms which are different from one another;
2. asymmetry may arise in molecules even when condition (1) is not fulfilled. Such asymmetry may be due to restricted rotations around double bonds or due to steric hindrance to free rotation around single bonds. A few examples are

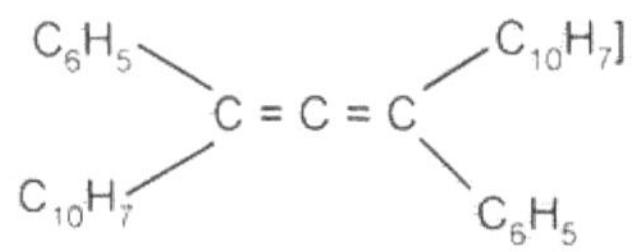

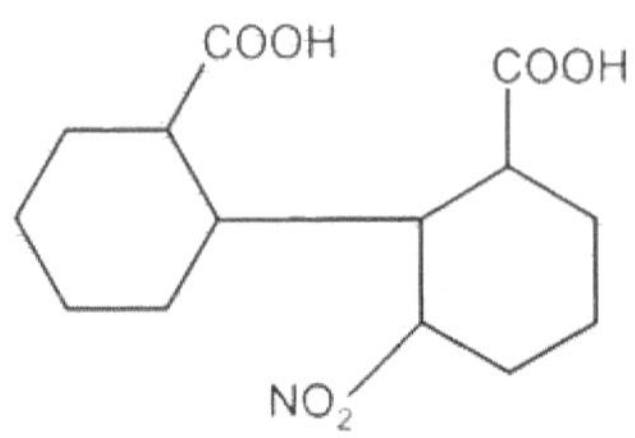

3. molecules have large helical structures, such as quartz.

Polarimeter Tube

The tube in which the solution or the liquid is placed for the examination of the rotatory power, generally consists of a thick walled tube with accurately ground ends, closed by means of circular glass plates with parallel sides. These plates are pressed against the ends of the tube by means of screw caps and rubber washers*. As the unit of length in polarimetery is a decimeter, the tubes are generally made in multiples of one decimeter. A two decimeter tube is more often used.

To maintain a constant temperature, the polarimeter tube may be enclosed in a jacket in which water at constant temperature can be circulated from a thermostat. Such jacketed tubes are used when very precise values are required.

AIM

To determine the specific rotation of sucrose at number of concentrations and obtain the value of intrinsic rotation for sucrose.

APPARATUS

1. Polarimeter 2. 100 ml flasks

CHEMICALS

1. Sucrose solution

THEORY

Ordinary light has vibration in all the possible directions whereas plane polarised light has vibration in only one direction.

Polaroids are those devices which allows only one vibration to pass through it. Hence light emerging out from a polaroid will be plane polarised (such as nicol polarising film, polaroid mirror, etc.).

In figure (a) an unpolarised light is passed through a polarising film. The emerging light is plane polarised. The second polaroid is placed with axis 90° rotated. No light will emerge out.

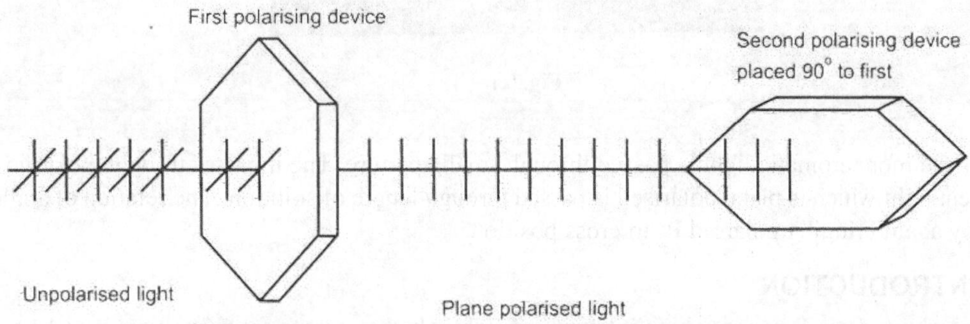

Fig. (a)

Certain substance like quartz sugar solution possess the property of rotating the plane of polarisation. These substances are called optically active substance. There are two types of substances. One which rotates the plane clockwise and the other anti-clockwise looking towards the direction of travel of light. If the rotation is anti-clockwise the substance is called dextro and if the rotation is clockwise the substance is called levo.

The sample solution under test is put in known length of tube. The tube is placed between 2 crissed polaroids as shown in Fig. (b)

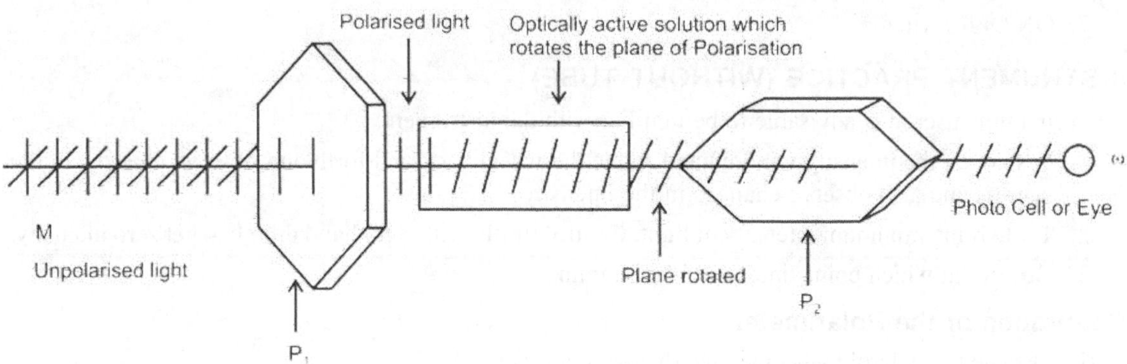

Fig. (b)

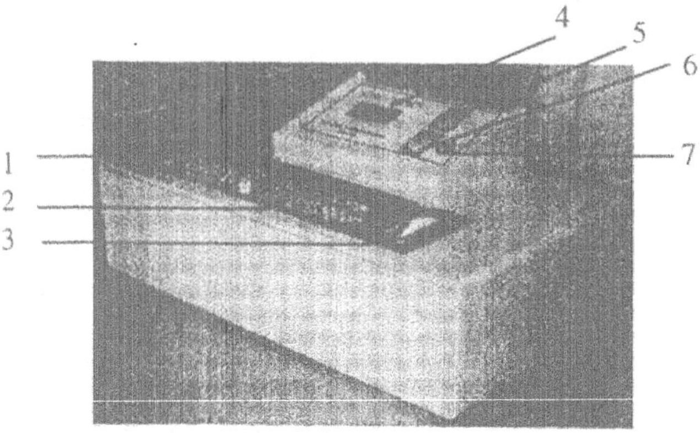

Fig. (c)

Unpolarised monocromatic light is passed through small aperture. The it passes through polaroid P_1. The emergent light which is plane polarised is passed through length of solution. The rotation of angle is measured by again bringing polaroid P_2 to cross position.

PANEL INTRODUCTION

1. *Intensity window:* Through which intensity of light which emerges from analyser is to be seen. The accuracy of result depends on how sharp min is caught. Try to obtain same pattern everytime.
2. *Sample chamber:* This is chamber for placement of sample to be studied. The chamber can accomodate max. length of 20 cms. of tube.
3. This is wheel to rotate analyser poloroid of the system. The position of this whell is related to degree meter.
4. *Digital degree meter:* It is capable of reading + 200° with accuracy of 1°.
5. 10 turn pot used to calibrate the degree scale with solution whose rotation is known. The shaft is to be rotated with sp, screw driver provided.
6. 10 T pot used to set degree meter to 'zero' with blank solution. The shaft is to be rotated with sp. screw driver.
7. ON/OFF switch.

INSTRUMENT PRACTICE (WITHOUT TUBE)

For a first time user, it is advisable to be familiar with the instrument.

1. Push open the intensity window and rotate the wheel back and forth and observe changes in the degress and also observe changes in the intensity.
2. Try to bring minimum intensity of light. Control firmly with right hand thumb to get zero intensity.
3. Observe at which points intensity is minimum.

Calibration of the Polarimeter

(i) Rotate the polaroid wheel so that display reads zero.

(ii) Fill the tube with distilled water and place it in the chamber.

(iii) Rotate the wheel back and forth so that intensity of light is minimum. At this position the display should read zero.If not, then with the help of screw driver adjust 'zero knob'm so that displat reads zero.

Now fill the tube with known (standard solution) and rotate the wheel in the direction of decreasing intensity, so that intensity of light is minimum. Adjust the calibrate pot so that displat reads + 26.4°.

(iv) Repeat steps 2 and 3 above.

Calibration Standard

20 gms. of AR sucrose when dissolved in distilled water to make 100 ml will give + 26.4 degrees

PROCEDURE

(i) Switch ON the instrument and allow it to warm up for about 5 minutes

(ii) Put plain water/solvent in the polarimeter tube. Care is to be taken that no bubble is trapped in the path. Any bubble which remains inside must be removed to centre bubble trap.

(iii) Rotate the polaroid wheel so that display reads ZERO.

(iv) Place the polarimeter tube inside the instrument and close the lid. Rotate the wheel back and forth so that intensity shows minimum. Note the readings of Degress say X_1. Do not disturb the set up.

(v) Now remove the tube and refill with 20% solution under test and insert it back in its position. The intensity will increase due to rotation of plane of polarisation. Rotate the wheel in the direction of decreasing intensity till intensity reaches minima. (Accurately control the rotation and arrive at exact minima). Note the reading on degree meter. Say X_2.

Repeat for various concentrations like 1%, 10%, 7.5%, 5%.

Angle of rotation (α) is

$$X_2 - X_1$$

In some cases where angle of rotation is more than 90° or 80° then it is difficult to decide which way plane has been turned. In this case 2 sample lengths must be used to arrive at direction and amount of rotation.

(vii) A graph is plotted between rotations and concentrations.

Note: Filter all the solution to remove dirt. Preferably use non-coloured substances for better results.

PRECAUTIONS

1. Always use AR/LR grade substance and filter if necessary
2. Make sure no bubble is trapped in the path of light.
3. While taking readings do not bend over the instrument. Also do not close on eye. Keep both the eyes open.
4. Wipe outside ends of tube body and glass with filter paper so that full clear view is obtanied.
5. At minimum position rotate the wheel back and forth to get exact minimum.
6. When the experiment is over, the solution or liquid used is removed from the polarimeter tube and the *tube is washed well with distilled water and dried before being stored.* The caps are unscrewed, or at least loosened, before storing.

OBSERVATIONS AND CALCULATIONS

No.	Concentration of solution (C)	Angle of rotation (α)	Specific rotation $[\alpha]_\lambda^t$
1	20%		
2	15%		
3	10%		
4	7.50%		
5	5%		

Mean =

Calculate the value of specific rotations by using the formula

$$[\alpha]_\lambda^t = \frac{100\,\alpha}{l\,c}$$

where

α = Angle of rotation

l = Length of polarimeter tube

c = Concentration of solution

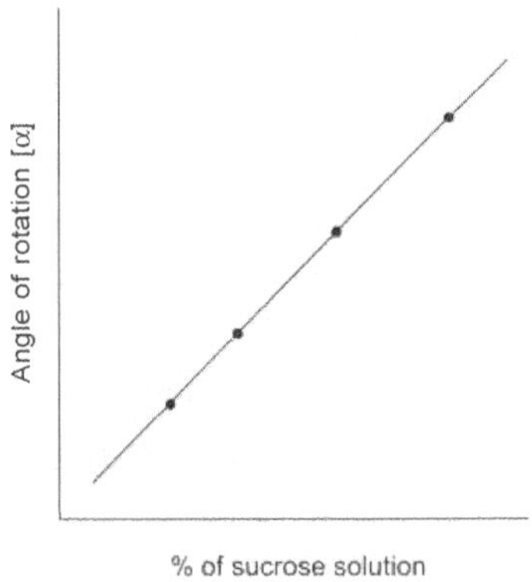

RESULT

The specific rotation of the sugar solution = ________ deg-dm^{-1}-g^{-1}.cm^3.

Intrinsic rotation is obtained from $[\alpha]_\lambda^t$ vs c graph extrapolated to $c = 0$.

EXPERIMENT – 24

Determination of Stability Constant by Job's Method

INTRODUCTION

Photometry is the measurement of intensity of light. When this measurement is made for light of particular colour or wavelength, it is called *colorimetry*.

Colorimetry is a method of analysis based on measuring intensities of colours of substances in solutions. This is usually done by comparing the intensity of colour of a solution with the colour of its solutions of known concentration (*standards*). Colour intensity measurements can also be made on the basis of absorption capacities of substances for some characteristic radiations. The latter techniques involves the use of photoelectric cells and are more often grouped under photoelectric colorimetry i.e., measurement of intensity of a coloured radiation through its photoelectric effect, if any.

Basic laws of colorimetry

Total intensity (I) of any light falling on a material body gets distributed into parts such as

Hints : Both these dyes absorb light in the visible range 400–630 mm. Use standard solutions in ethanol.

Crystal violet	0.0056 gL^{-1} 7 (Mol. wt. : 408)
Aurine	0.038 g L^{-1} (Mol. wt. : 290)

Make a mixture by mixing equal volumes of the two solutions and another by mixing them in any other volume ratio. Carry out the measurements and compare the results obtained with those already known.

AIM

To find the composition of ferric ions-salicylic acid complex absorptiometrically, using Job's method.

APPARATUS

1. Absorptiometer

CHEMICALS

1. 0.002M HCl
2. 0.001M salicylic acid in 0.002 M HCl
3. 0.001M Fe^{3+} ions as ferric alum in 0.002 M HCl.

THEORY

This complex is stable in acid solutions in pH range 2.6 to 2.8. This range can be obtained by using solutions 0.002 M in HCl. At this pH value, the phenolic –OH and –COOH group of the acid remain unionised. Since the complex is coloured, its concentrations can be measured on an ordinary absorptiometer operating with white light and coloured glass filters.

In any complex formation, the bonding units combine in integral numbers.

$$xFe^{3+} + yC_6H_4(OH)COOH \rightleftharpoons Complex$$

According to Job's method, when equimolar solutions of two reactants are mixed in varying proportions, at equilibrium the maximum amount of the complex is formed when the two reactants are present in the same mole ratio as is required for the complex formation.

When the complex is coloured and the reactants are colourless or when we have a filter which acts as absorbent for the complex and not for the reactants, the equilibrium concentration of the complex can be followed by measuring optical density of the solution. In the present case, the complex formed is deep violet in colour. It wilt be quite easy to choose a suitable filter for it and find out the volume ratio of the two solutions for which optical density of the solution is maximum (Fig. 1).

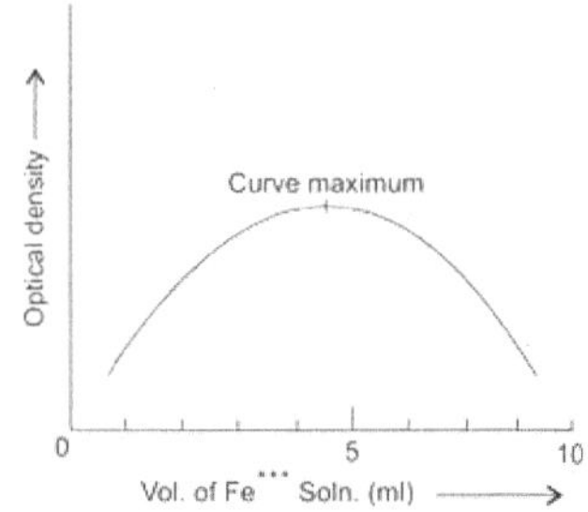

Fig. 1

PROCEDURE

(i) Make the following mixed solution:

Ferric ion solution (ml)	9	8	7	6	5	4	3	2	1
Salicylic acid solution (ml)	1	2	3	4	5	6	7	8	9

(ii) Take one of these solutions and choose the filter for maximum absorption of light. Record the wavelength of light transmitted by this filter and the number of the filter if any. Measure transmittance and optical density for each of the above solutions. Plot transmittance (or optical density) against volume of Fe^{3+} ions solution in the mixtures. Mark the minimum (or maximum) point on the curve. This corresponds to the composition of the complex. In this case, it will be found to be for a molar ratio of 1:1 for Fe^{3+} ions and salicylic acid molecules.

(iii) The above data can be further used for finding the equilibrium, disassociation or stability constant of the complex. We shall need a calibration curve of optical densities of the complex solutions versus concentrations of complex. For this purpose, we prepare a number of dilutions of ferric alum solution in 0.002M HCl. Convert the ferric ions in each solution practically completely to the complex form by saturation with salicyclic acid powder. This gives the concentration of the complex in each solution. Now measure the optical densities and draw the optical density versus concentrations curve for the coloured complex.

(iv) For each of the solutions under the first experiment, the observed optical density corresponds to concentration of the complex. Since composition of the complex is known the residual concentrations of the ferric ions and salicylic acid in the equilibrium mixtures can be calculated. Do this for three or four solutions.

OBSERVATIONS AND CALCULATIONS

Solution No.	Concentration (moles/litre)		
	Fe^{3+}	$C_6H_4(OH)COOH$	Complex
1			
2			
3			
4			

Equilibrium constant for the formation reaction

$$Fe^{3+} + C_6H_4(OH)\,COOH \rightleftharpoons [Fe.C_6H_4(OH)COOH]^{3+}$$

can be obtained from

$$K = \frac{\left[\left[Fe\,(C_6H_4(OH)\,COOH)\right]^{3+}\right]}{\left[Fe^{3+}\right]\left[C_6H_4(OH)COOH\right]}$$

From this value of K_C, ΔG for the formation reaction in 1M solution can be obtained.

$$\Delta G = -RT \quad ln\ K_C$$

RESULT

The stability constant of the iron salicylic acid complex is found to be (K) = ———

EXPERIMENT - 25 Determination of Fluoride Ion using an Ion Selective Electrode

INTRODUCTION

In recent years direct potentiometry has become important as an analytical technique largely because of the development of ion-selective electrodes (ISE). This type of electrode incorporates a special ion-sensitive membrane which may be glass, a crystalline inorganic material or an organic ion-exchanger. The membrane interacts specifically with the ion of choice, in our case fluoride, allowing the electrical potential of the half cell to be controlled predominantly by the F- concentration.

The potential of the ISE is measured against a suitable reference electrode using an electrometer or pH meter. The electrode potential is related to the logarithm of the concentration of the measured ion by the Nernst equation.

$$E = E^{0} + 2.303\frac{RT}{nF}\log[M]$$

where n is the ion charge (negative for anions). The factor 2.303 RT/F has a theoretical value of 59 mV at 25 °C. The equation is only valid for very dilute solutions or for solutions where the ionic strength is constant. Ionic strength is defined by

$$I = \frac{1}{2}\sum Z_i^2 C_i$$

where Z_i is the charge on an ion and C_i is its concentration. ISEs are available for measuring more than 20 different cations (e.g., Ag^+, Na^+, K^+, Ca^{++}, Cu^{++}) and anions (e.g., F^-, Cl^-, S^{-2}, CN^-).

In this experiment you will use a fluoride sensitive electrode and either a saturated calomel electrode (SCE) or Ag/AgCl external reference electrode to measure the fluoride-ion content of a solution. Fluoride is added to drinking water and toothpaste to inhibit dental caries; it is also present in effluents from many industrial processes, e.g., manufacture of fluoro-polymers. Flouride ISEs only respond to free ionized F- in solution and can thus be used to measure this ion in the presence of other fluorine compounds, e.g., AlF_6^{3-} or organofluorine compounds. In other words, the electrode responds to F activity.

AIM

To determine the fluoride ion in the given unknown solution using Ion selective electrode.

APPARATUS

1. Fluoride ISE and Ag/AgCl or SCE reference electrode.
2. Multimeter or pH meter capable of displaying mV potentials
3. Beakers
4. Volumeteric flask
5. Glass rod

CHEMICALS

1. NaF, dried at 100 °C for 1 hour.
2. Liquid NaF unknown.
3. KCI (7.55 g)

PROCEDURE

Preparation of Standards

(i) Dry the NaF solid for 1 hour at 100 °C.

(ii) Accurately weigh out about 0.42 g of NaF, dissolve in deionized water, dilute to 100 ml in a volumetric flask and mix well. This solution is about 10^{-1} F in NaF.

(iii) Transfer 10.0 ml of the solution prepared in (2) to a 100 ml volumetric flask using a pippet, dilute to volume with deionized water and mix well. This solution is about 10^{-2}F in NaF.

(iv) Weigh out 7.55 g of KCl on a top-loading balance and dissolve in 100 ml of deionized water. This solution is 1 F in KC1.

(v) Prepare standard solutions in four 100 ml volumetric flasks as follows:

	mL 10^{-2} F NaF (from 3)	ml 1 F KC1
(I)	1.00	10.00
(II)	2.00	10.00
(III)	5.00	10.00
(IV)	10.00	10.00

Dilute each flask to volume with deionized water and mix well.

Instrument Setup and Operation

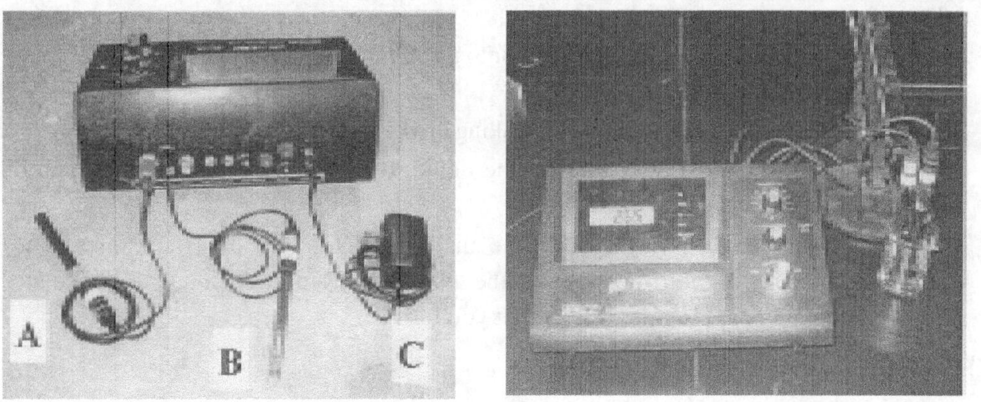

Fig. 1. Instrument setup for ISE measurements. *Left:* Back of pH/mV meter showing ISE (A), reference electrode (B) and power (C) connections. *Right:* Front view showing typical arrangement for mV measurements. Note that electrodes are suspended off the bottom of the beaker that holds the sample.

Note: When the electrodes are not immersed in solution, be sure to set the meter to the "OFF" position to avoid polarizing the electrodes. During measurement, set the meter to the "mV" position, making sure that the "mV" indicator light is illuminated.

Analysis of Unknown

Your unknown for this experiment is a solution. When you obtain your unknown, you need to quantitatively transfer it to a 100 ml volumetric flask and dilute it to the mark, resulting in the "prepared" unknown solution. You are to report the results of this "prepared" unknown.

(i) Add 1 ml of your prepared unknown, then 10 ml of KCl to a 100 ml volumetric flask. Dilute to the 100 ml mark with deionized water.

(ii) Measure the potential in mV of the fluoride ISE vs the reference electrode for each of the four standards and unknowns.

Caution: Do not touch the ISE ion sensitive membrane. Rinse it with deionized water between measurements and then with a small volume of the new solution. Do not wipe it dry.

(iii) Pour about 30 ml of each standard or unknown solution into a clean, dry 100 ml beaker and immerse the electrodes in the solution to a depth of not more than 2 cm, as shown in Figure 1. Measure the electrode potential, taking care to note both the sign and the magnitude of the potential.

(iv) When you finish, rinse the electrodes with deionized water. Leave the reference electrode in the appropriate storage solution. The F^- ISE should be stored dry and loosely capped. DO NOT force the cap onto the electrode tip!

Analysis of Fluoride in Toothpaste

The toothpaste sample should be prepared and analyzed at the same time the standards and your prepared unknown.

(i) Accurately weigh about 0.2 g of toothpaste into a 100 ml beaker. Add 10 ml of 1F KCl and about 40 mL of water to the beaker.

(ii) Boil the mixture gently for 3-5 minutes, breaking up the toothpaste with a stirring rod if necessary.

(iii) Cool the solution, quantitatively transfer the liquid to a 100 ml volumetric flask and dilute to volume.

(iv) Analyze this sample when you perform the analysis on your standards and unknown. Report your results in terms of the % (w/w) F^- in the toothpaste. Report this on your unknown card in addition to the unknown sample results that you turn in.

OBSERVATIONS AND CALCULATIONS

1. Accurately calculate the molarity of NaF for each of the standard solutions.
2. Plot a graph of the logarithm of [NaF] in the standard solutions vs E. Determine the best line through the four standard solution experimental points. Calculate the slope of the calibration curve, and its associated uncertainty.

   Slope = dE/dlog [NaF].

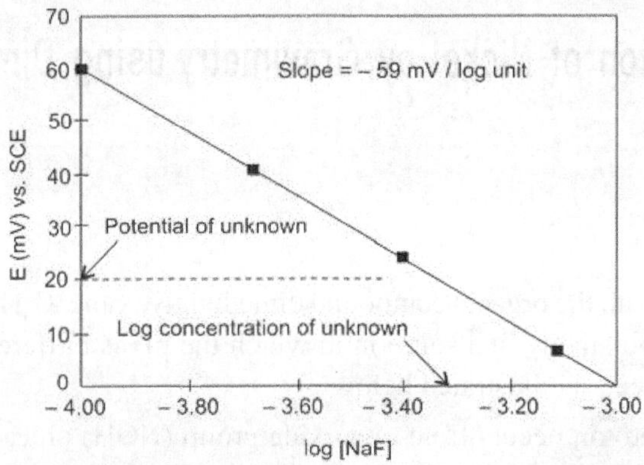

Fig. 2 Calibration Curve for Fluoride Electrode

3. Utilizing the calibration curve, determine the concentration of NaF in your "prepared" unknown solution. Report this as percent fluoride (% wlv) in the "prepared" unknown. Report the 95% confidence interval for your results.

RESULT

The concentration of the F^- ion solution = ——— moles/litre

Experiment – 26 Estimation of Nickel by Gravimetry using Dimethylglyoxime

INTRODUCTION

Nickel(II) forms a precipitate with the organic compound dimethylglyoxime, $C_4H_6(NOH)_2$. The formation of the red chelate occurs quantitatively in a solution in which the pH is buffered in the range of 5 to 9. The chelation reaction that occurs is illustrated below.

Although the loss of one proton occurs from one oxide group (NOH) on each of the two molecules of dimethylglyoxime, the chelation reaction occurs due to donation of the electron pairs on the four

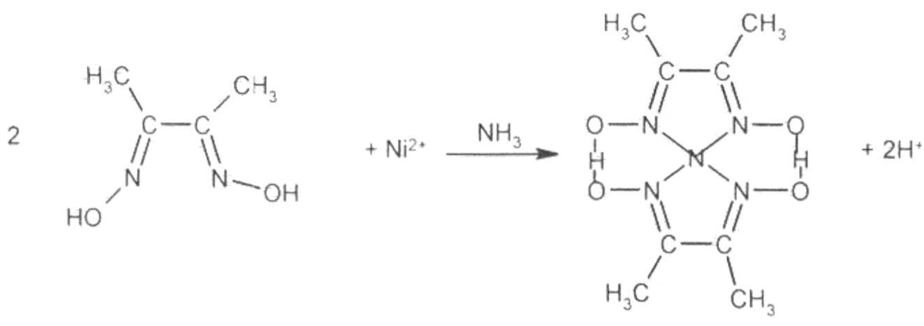

Dimethylglyoxime (DMG) Ni(DMG)$_2$

nitrogen atoms, not by electrons on the oxygen atoms. The reaction is performed in a solution buffered by either an ammonia or citrate buffer to prevent the pH of the solution from falling below 5. If the pH does become too low the equilibrium of the above reaction favors the formation of the nickel(II) ion, causing the dissolution of $Ni(DMG)_2$ back into the mother liquor.

Adding tartarate or citrate ions before the precipitation of the red nickel complex prevents interference from Cr, Fe and other metals. These anions selectively form tightly bound soluble complexes with the metals and prevent the formation of insoluble metal hydroxides in the buffered solution.

An alcoholic solution of dimethyglyoxime (DMG) is used as the precipitating reagent during the experiment because DMG is only slightly soluble in water (0.063 g in 100 ml at 25°C). It is therefore crucial to avoid the addition of too large an excess of the reagent because it may crystallize out with the chelate. It is also important to know that the complex itself is slightly soluble to some extent in alcoholic solutions. By keeping the volume added of the chelating reagent small, the errors from these sources are minimized. The amount of the reagent added is also governed by the presence of other metals such as cobalt, which form soluble complexes with the reagent. If a high quantity of these ions is present, a greater amount of DMG must be added. The nickel dimethyiglyoximate is a precipitate that is very bulky in character. Therefore, the sample weight used in the analysis must be carefully controlled to allow more convenient handling of the precipitate during transferral to the filtering crucible. To improve the compactness of the precipitate, homogeneous precipitation is often performed in the analytical scheme.

This is accomplished by the adjustment of the pH to 3 or 4, followed by the addition of urea. The solution is heated to cause the generation of ammonia by the hydrolysis of the added urea, as indicated by the following reaction.

$$NH_2CONH_2 + H_2O = 2\,NH_3 + CO_2$$

A slow increase in the concentration of ammonia in the solution causes the pH to rise slowly and results in the gradual precipitation of the complex. The result is the formation of a more dense, easily handled precipitate. Once the filtrate has been collected and dried, the nickel content of the solution is calculated stoichiometrically from the weight of the precipitate.

AIM

To determine the percentage of nickel in a given sample of nickel-steel by employing dimethylglyoxime

APPARATUS

1. 3-Sintered Glass Crucibles: medium porosity (see instructor for these)
2. 3-400 ml beakers
3. 3-Watchglasses and glass hooks
4. 1-Rubber Policeman and glass stirring rod
5. Whatman No. 40 filter paper; medium porosity

CHEMICALS

1. Unknown Ni ore sample
2. Nitric Acid (concentrated)
3. Hydrochloric Acid (concentrated)
4. 20% Tartaric Acid solution (prepared by student)
5. 1:1 Ammonium Hydroxide solution (prepared by student)
6. 1% alcoholic dimethyiglyoxime solution
7. Urea (solid, ACS reagent grade)
8. Acidic $AgNO_3$ (already prepared)
9. pH paper

THEORY

The precipitation of nickel as nickel dimethylglyoximate is the easiest and quickest method of estimating nickel in its compounds. In this method, nickel is precipitated by the addition of an alcoholic solution of dimethylglyoxime to a warm, faintly acidic solution of a nickel salt, followed by the addition of a slight excess of NH_4OH. Dimethylglyoxime is a good *chelating agent* and provides a selective method for the determination of nickel.

The object of this experiment is to give some training to students in the quantitative separation of an element (nickel) from the other elements present in steel and in the use of chelating agents in quantitative analysis.

PROCEDURE

(i) Weight about one gram of the steel sample to the fourth significant figure and put it in a 250 ml beaker.

(ii) Add 20–25 ml of concentrated hydrochloric acid and warm it if the steel is not completely dissolved.

(iii) Heat the solution to boiling and add 10 ml of concentrated nitric acid to ensure complete oxidation of iron from the ferrous to the ferric state.

(iv) Dilute it to about 70–80 ml and filter it to remove any solid material which is left over undissolved. Wash the filter paper with hot water and collect the filterate and washings in a 400 ml beaker.

(v) Dilute it to about 250 ml with distilled water. Add 7 to 8 g of tartaric acid and neutralize most of the acid of the solution by adding ammonium hydroxide. The solution should be barely acidic as tested by a litmus paper.

(vi) Heat the solution to 80°C and add 1% alcohol solution of dimethylglyoxime in slight excess (about 30 ml).

(vii) Immediately add dilute ammonia dropwise, with constant stirring, until the solution is slightly alkaline.

(viii) Stir well, and allow it to digest on a water bath for half an hour. In the mean time, heat a clean sintered glass crucible at 120°C in an air oven (or prepare a Goooch crucible) and find its constant weight.

(ix) Filter off the pink colored precipitate through the weighed crucible and wash the precipitate with cold water until it is free from chloride (as tested by taking a few ml of the filtrate and adding a few drops of nitric acid, followed by one ml of silver nitrate solution).

(x) Dry the precipitate at 110°C in an air oven for one hour to a constant weigh and weight nickel as $N_1(C_4H_7O_2N_2)_2$.

OBSERVATIONS AND CALCULATIONS

Weight of watch glass + steel (1) = a g

Weight of watch glass + steel (ii)	= b	g
Weight of steel taken	= (a – b)	g
Weight of the sintered glass or the Gooch crucible	= Z	g
Weight of the sintered glass or the Gooch crucible + nickel precipitate	= y	g
Weight of nickel dimethylglyoximate	= (y – z)	g
Weight of nickel in the precipitate (1)	= (y – z) × 0.203	

% nickel in the steel $= \frac{(y - z) \times 0.203 \times 100}{(a - b)}$

RESULT

% nickel in the steel = ____________

Part - II Physical Chemistry Experiments

CONDUCTOMETRY

INTRODUCTION

Conductometry is the measurement of conductivity of a solution due to the mobility of cations and anions towards respective electrodes. Conductivity is inversely proportional to resistance of a solution C = 1/R. The unit of conductivity is mhos.Conductivity of a solution depends upon number of ions (conc.), charge of ions, size of ions and temperature. All the laws which are applicable to solid conductor's substances also can be applied to the solution containing ions. Accordingly, the resistance of solution is given by

$$R = E/I$$

where E = potential difference and I = current which flows through

The unit of R is ohms. potential difference is (E) volts and that of current is amperes. The R of a solution depends upon the length and cross section of the conductor through which conductivity takes place, therefore

$$R = \rho l/a$$

Where ρ = Specific résistance, hence

$$\rho = aR/l$$

***Specific résistance* (ρ)** is the resistance offered by a substance of 1 cm length and 1 sq.cm surface area. Unit of measurement is ohm cm.

***Specific conductivity* (k_l)** is the conductivity offered by a substance one cm length and 1sq.cm surface area. Unit of measurement is mhos cm^{-1}.

$$k_v = 1/\rho$$

***Equivalent conductivity*(λ_v):** Is the conductivity of a solution containing equivalent weight of the solute between electrodes 1cm apart, 1sq.cm surface area. Unit of measurement is mhos cm^{-1}.

Equivalent conductivity = specific conductivity × volume of solution containing 1 gm equivalent weight of electrolyte

***Molar conductivity* (μ_v)** is the conductivity of a solution containing molecular weight of solute between electrodes 1cm apart, 1sq.cm surface area.

Molar conductivity = specific conductivity × volume of solution containing one molecule weight of the electrolyte

The determination of end point of a titration by means of conductivity measurements is known as conductometric titration. During the course of a titration, the conductivity of solution changes, since there is change in the number and mobility of ions. At the end point of titration, there is a sharp change in the conductivity of a solution shown by the intersection of the lines in the graph of conductivity Vs volume of titrant added.

Advantages of conductometric titrations

1. Determination of specific conductivity is not required.
2. It is not necessary to use conductivity water.
3. No indicator is necessary.
4. Titrations can be done with coloured or dilute solutions or turbid suspensions.
5. Incompletion at the end point does not effect the results as few measurements before and after the end point are sufficient.
6. The principle that conductivity depends upon the no. and mobility of the ions is used and curves are mostly straight lines. Hence few measurements are sufficient.
7. As the end point is determined graphically, errors are minimised and accurate end point can be determined.
8. The cell constant need not be determined, provided the same electrode is used throughout the experiment.
9. Temperature need not be known, provided it is maintained constant throughout the titration.

CELL CONSTANT AND ITS DETERMINATION

Theory

If 'c' is the conductance of a solution measured by a cell with electrode of cross sectional area 'a' cm^2 and 'l' cm apart, the specific conductivity of the solution is given by

$$k = c \times \frac{l}{a}$$

where $$\frac{l}{a} = \frac{k}{c} \qquad \left(\because \frac{l}{a} = k\right)$$

The cell constant i.e., the ratio l/a cannot be obtained from the geometrical dimensions of the cell for both 'l' and 'a' are not accurately known. It is therefore necessary to calibrate the cell with a solution of known specific conductivity. Usually 0.1 or 0.01 N Kcl solution is used

$$\text{(k) cell constant} = \frac{\text{Specific Conductivity}}{\text{Conductance}}$$

Procedure

Prepare 0.1M KCl solution and transfer the solution in 100 ml beaker. Wash the conductivity cell with distilled water and wipe (electrode) with tissue paper. Dip the electrode in the kc$l$ solution taken in a beaker and measure the conductance. Note the readings.

Calculations

Specific Conductivity of KCl

Units: $Ohm^{-1}\ cm^{-1}$

0.1 N specific conductance is 0.01224

Similarly

0.01 N specific conductance is 0.001412 at 25°C

$$0.1 = \frac{wt(x)}{74.5} \times \frac{1000}{100}$$

To prepare 0.1 N Kcl in 100 ml the wt (x) is

$$x = \frac{0.1 \times 74.5}{10} = 0.745 \text{ g}$$

To prepare 0.01 N Kc*l* in 100 m*l* from the above prepared solution

$$M_1V_1 = M_2V_2$$

$$0.1 \times V_1 = 0.01 \times 100$$

$$V_1 = \frac{0.01 \times 100}{0.1} = 10 \text{ m}l \text{ (10 ml of the stock solution has to be diluted to 100 ml)}$$

$$10^6 \ \mu℧ = 1℧$$

$$1\mu℧ = 10^{-6}℧$$

$$\text{Conductivity} = 13.44 \ \mu℧$$

$$= 13.44 \times 1000$$

$$= 13440 \ \mu℧ = 13440 \times 10^{-6} ℧$$

$$k = \frac{0.01224}{0.01344} = 0.91 \text{ cm}^{-1}$$

METHOD OF CALIBRATION

1. **Approx. method (accuracy 2% to 3%) for cell const. k = 1 only.** Proceed as given below.
   (a) Allow the instrument to warm up for 15 minutes
   (b) Bring in standard conductance of 1.00 mM by throwing the switch down
   (c) With the help of screwdriver turn the standardize shaft till the Digital display reads 1.000
   (d) Throw the standard cond. switch up

   The instrument is now ready for use.

2. **Accurate Method (0.5% to 1% accuracy)**
   (a) Prepare KC*l* Sol. and allow it to attain the room / desired temperature. The solution so prepared is temperature dependant
   (b) Dip the cell k = 1 in the solution
   (c) Check that the std. cond. switch is in upward position.
   (d) Turn the standardise shaft with screwdriver till the display reads the correct conductance of the solution. The temp. effect has to be considered.

The instrument thus calibrated also accounts for deviation in cell constant to the extent of ± 10%. k = 0.5 can also be used in this method.

Conductivity (in milli m mho/cm)

Temp °C	KCl 1N	KCl 0.1N	KCl 0.02N	KCl 0.01N
20	0102.07	0011.67	0002.501	0001.278
21	0104.00	0011.91	0002.553	0001.305
22	0105.94	0012.15	0002.606	0001.332
23	0107.89	0012.39	0002.659	0001.359
24	0109.84	0012.64	0002.712	0001.386
25	0111.80	0012.88	0002.765	0001.413
26	0113.77	0013.13	0002.819	0001.441
27	0115.74	0013.37	0002.873	0001.468
28	-	0013.62	0002.927	0001.496
29	-	0013.87	0002.981	0001.524
30	-	0014.12	0003.036	0001.552
31	-	0014.37	0003.091	0001.581
32	-	0014.62	0003.146	0001.609
33	-	0014.88	0003.201	0001.638
34	-	0015.13	0003.256	0001.667
35	-	0015.39	0003.312	

TEMPERATURE COMPENSATION

The conductance of a solution varies sharply with temperature. Hence compensation is very important.

The standard practice is to refer to the conductance at 25°C. Hence it is desirable to compute measurement of various temp. to 25°C. Temperature compensation has therefore to be applied at other temp. at 2% per C. However this is approximation. the correction factor varies with substance, temp., dilution, etc. Hence accurate compensation is very difficult.

CONDUCTIVITY CELL

It consists of two platinum plates firmly placed in an insulating container which serves to isolate a portion of the liquid. The electrodes are coated with spongy black platinum. which increases the effective surface and reduces the polarizing surface. This coating can be easily deposited and is resistant to mechanical or chemical abrasion.

MAINTENANCE OF CONDUCTIVITY CELL

Before using the new conductivity cell it should be soaked in distilled water for 24 hours. Also after use the cell should be thoroughly cleaned in water and stored in distilled water. If stored dry. it should be soaked in distilled water for 2 hours before use, to obtain dependable readings.

PRECAUTIONS

1. The conductivity cell, when not in use, should always be kept immersed in distilled water to prevent drying of the platinum black electrode. If it gets dry, keep the cell immersed under hot water at about 50°C for few hours. Remove air bubbles, if any, that stick to the platinum plates.
2. When the plates of conductivity cell get dirty, clean them in dilute potassium-dichromate in H_2SO_4 for one day and then rinse in running water followed by rinsing in distilled water.

RELATION BETWEEN CONDUCTIVITY & RESISTIVITY

Conductivity is reciprocal of Resistivity and is expressed in Mhows or Siemens.

Thus $C = \frac{1}{R}$

where C is in Mhos

and R is in ohms

1 Mho = 1 Siemens

EXPERIMENT - 27 Conductometric Titration of Strong Acid Vs Strong Base

AIM

To determine the end point of the titration of strong acid against strong base by conductometrically.

APPARATUS

1. Conductivity meter
2. Conductivity cell
3. Micro burette
4. Beaker
5. Volumetric flasks
6. Glass rod

CHEMICALS

1. 0.1 N HCl
2. 0.1N KCl
3. 1 N Oxallic Acid
4. 1 N NaOH

THEORY

The determination of equivalence point of a titration conductometrically is based upon the measurement of the conductance during the course of titration which varies in different manner before and after the equivalence point. Consider a titration of strong acid like Hydrochloric acid Vs Sodium hydroxide. The equation will be

$$HCl + NaOH \rightarrow NaCl + H_2O$$

When HCl is taken in beaker as titrate, the initial conductivity is high, because strong acid completely dissociates into H^+ ions and the ionic conductivity of H^+ is 350. When NaOH is added as titrant, the OH^- and H^+ reacts to produce water and the no. of H^+ decreases and the conductivity gradually decreases after every addition. After the end point, when all the H^+ has reacted, the addition of NaOH causes increase in the no. of OH^- and hence the conductivity starts to increase (ionic conductivity of OH^- is 199). A plot conductivity Vs volume of NaOH added will consist of two straight line branches intersecting at the neutralization point like V shape as shown in below graph.

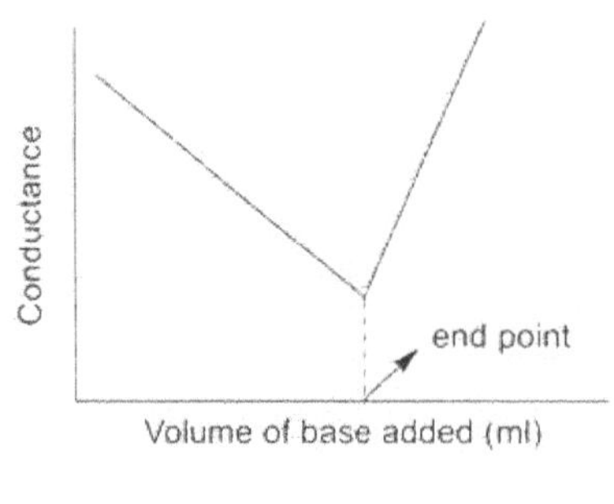

PREPARATION OF REAGENTS

1. ***0.1 N HCl (approx):*** Dilute 0.86 ml of 11.6 N and 36 % concentrated HCl (Mol. Wt. 36.5) to 100 ml with distilled water.
2. ***0.1N KCl:*** Dissolve 7.456 gm of KCl (Mol. Wt. 74.56) (chemically pure) in 1 litre of double distilled water.

3. ***1 N Oxallic Acid:*** Dissolve 63.04 gm of oxalic acid (Mol. Wt. 126.08; Eq. Wt.63.04) in 1 litre of distilled water.
4. ***1 N NaOH:*** Dissolve 10 gm of NaOH (Mol. Wt. 40) in 250 ml of distilled water.

PROCEDURE

(a) ***Standardisation of NaOH Solution:*** Pipette out 20 ml of prepared NaOH solution in a clean conical flask. Add 1 or 2 drops of methyl orange indicator and titrate against standard oxalic acid solution. Note down the end point at which the color changes from pale yellow to pale pink.

(b) ***Determination of Cell Constant:*** Take a conductivity cell and determine its cell constant using a 0.1N KCl Solution (procedure given in introduction of Conductometry)

(c) ***Conductometric titration***

(i) Fill the burette with standard 1N NaOH solution

(ii) Take 25 ml of the given HCl solution in a 100 ml beaker and the dip the conductivity cell in it and measure the conductance initially.

(iii) Now add NaOH from burette dropwise, i.e. 0.1 ml for each of the addition. After each of the addition, stir the solution gently by shaking and note down the change in conductance.

(iv) The measured conductance are recorded and tabulated in the table.

(v) Plot the graph between conductivity against volume of base added. the intersection of two straight lines gives the end point as shown in the above graph.

(vi) Calculate the strength of the given strong acid (HCl) from the known strength of the given NaOH solution.

OBSERVATIONS AND CALCULATIONS

The titre value corresponding to the point of inflection in the end point graph is—ml

Therefore

$$\text{Strength of HCl} = \frac{\text{End point titre value} \times \text{Normality of NaOH (A)}}{\text{Volume of HCl taken in breaker}} = \frac{\text{End point volume ml} \times \text{(A)}}{25\text{ ml}}$$

$$= \_\_\_\_\_\_\_\_\ \text{N}$$

Standardization NaOH Vs Strong Acid

Sr.No.	Volume of base added (ml)	Conductance (siemens)

RESULT

The end point for measured conductance with volume of alkali (ml) curve is ——— ml

The strength of hydrochloric acid calculated is ——— N

EXPERIMENT - 28 Conductometric Titration of Mixture of Acids VS Strong Base

AIM

To determine the composition of mixture of acetic and HCl acids by conductometric titration.

APPARATUS

1. Conductivity meter
2. Conductivity cell
3. Micro burette
4. Beaker
5. Volumetric flasks
6. Glass rod

CHEMICALS

1. 0.1N Acetic acid (CH_3COOH)
2. 0.1N Hydrochloric acid (HCl)
3. 1N Oxallic acid
4. 0.1 N KCl
5. 1N sodium hydroxide (NaOH)

THEORY

When a mixture containing acetic acid and Hydrochloric acid is titrated against an alkaline, strong acid (HCl) will be neutralized first. The neutralization of the weak acid (CH_3COOH) commences only after the complete neutralization of strong acid. Thus the conductance titration curve will be marked by two breaks. The first one corresponds to the equivalence point of Hydrochloric acid and the second to that of acetic acid.

Let V_1 and V_2 ml be the volumes of alkaline corresponding to first and second breaks respectively. Then V_1 ml of sodium hydroxide (NaOH) equivalent to Hydrochloric acid (HCl). $V_2 - V_1$ ml of sodium hydroxide equivalent to acetic acid (CH_3COOH).

$$HCl + NaOH \rightarrow NaCl + H_2O$$

$$CH_3COOH + NaOH \rightarrow CH_3COONa + H_2O$$

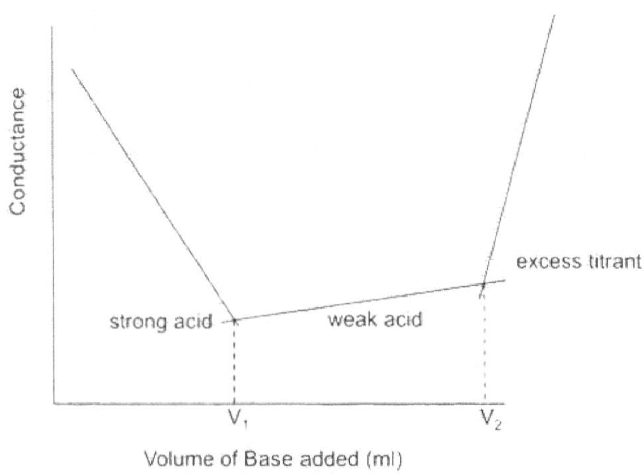

PREPARATION OF REAGENTS

1. ***0.1 N HCl (approx):*** Dilute 0.86 ml of ll.6 N and 36 % concentrated HCl (Mol. Wt. 36.5) to 100 ml with distilled water.
2. ***0.1 N Acetic acid (approx):*** Dilute 5.74 ml of 17.4 N and 99.5% acetic acid to 1litre with distilled water.
3. ***1 N Oxallic Acid:*** Dissolve 63.04 gm of oxalic acid (Mol. Wt. 126.08; Eq. Wt.63.04) in 1 litre of distilled water.
4. ***0.1N KCl:*** Dissolve 7.456 gm of KCl (Mol. Wt. 74.56) (chemically pure) in 1 litre of double distilled water.
5. ***1 N NaOH:*** Dissolve 10 gm of NaOH (Mol. Wt. 40) in 250 ml of distilled water.

PROCEDURE

(a) ***Standardisation of NaOH Solution:*** Pipette out 20 ml of prepared NaOH solution in a clean conical flask. Add 1 or 2 drops of methyl orange indicator and titrate against standard oxallic acid solution. Note down the end point at which the color changes from pale yellow to pale pink. Calculate the normality of NaOH solution (A).

(b) ***Determination of Cell Constant:*** Take a conductivity cell and determine its cell constant using a 0.1N KCl Solution (procedure given in introduction of Conductometry).

(c) ***Conductometric titration:***

(i) Fill the burette with standard 1N NaOH solution.

(ii) Take the given mixture (25 ml of 0.1N CH_3COOH + 25 ml of 0.1N HCl) in a small beaker and titrate it against 1N sodium hydroxide conductometrically.

(iii) Add 0.1ml of sodium hydroxide from micro burette and stir well and note down the conductance of the solution.

(iv) Repeat the determination by stirring well after each addition of 0.1ml of sodium hydroxide (NaOH) until the two end points are obtained.

(v) Plot a graph volume of NaOH Vs conductance.

OBSERVATIONS AND CALCULATIONS

The volume of NaOH corresponding to the first point of inflection is $V_1 = X_1$ ml and that corresponding to the second point of inflection is $V_2 = X_2$ ml.

$$V_2 - V_1 = X_3 \text{ ml}$$

Hence the HCl and CH_3COOH in their mixture of volume (50 ml) consume respectively

$$X_1 \text{ ml and } X_3 \text{ ml}$$

Therefore

$$\text{Strength of Strong acid (HCl)} = \frac{X_1 \text{ ml} \times \text{conc. of NaOH}}{25 \text{ ml}} = \_\_\_\_\_ \text{N}$$

$$\text{Strength of Weak acid } (CH_3COOH) = \frac{X_3 \text{ ml} \times \text{conc. of NaOH (A)}}{25 \text{ ml}} = \_\_\_\_\_ \text{N}$$

Standardization NaOH Vs Mixture of Acids

Volume of base added (ml)	Conductance (Siemens)

RESULT

The normality of acetic acid and hydrochloric acid (HCl) are = ——— N and ——— N respectively.

POTENTIOMETRY

INTRODUCTION

The variation of potential of an electrode with the .concentration of ions with which it is in equilibrium may be used as an indicator in volumetric analysis. The method is applicable to wide range of titrations, provided that an appropriate electrode (indicator electrode) is available. An indicator electrode is the one whose potential indicates the change in the concentration of the ions to be titrated. As it is not possible to determine the electrode potential separately, the indicator electrode is used in conjuction with a reference electrode, the potential of which remains constant during the course of titration. Most commonly used reference electrode is the saturated calomel electrode (SCE). Suppose a solution of an acid is titrated with a solution of an alkali. The following cell is set up in the acid solution

H_2(Pt) | acid solution || KCl aq. || calomel electrode

The e. m f. of such a cell is given by

$$E = E_{cal} - E_{H} = E' + 0.05916\ pH \qquad (at\ 25^{o}C)$$

In carrying out the titration, the reagent is added in small amounts and the mixture is stirred thoroughly. In order to obtain the graph in the neighbourhood of equivalence point with greater precision, the titrant must be added in smaller and smaller amounts as the equivalence point reaches.

ADVANTAGES

1. Potentiometric titrations are applicable to wide range of reactions, provided an appropriate indicator electrode is available. The method is of particular use in the titration of highly coloured solutions where indicators can not be used.
2. The determination of equivalence point by this method depends upon a series of independent observations rather than on one judgement. The method, thus, gives very reliable results.
3. Extremely dilute solutions may also be titrated by this method.
4. It is possible to carry out a stepwise titration of polybasic acids and mixtures of acids. Each step has its own inflexion point.

Acid base titrations: Acid-base titrations can be carried out with either hydrogen electrode, or quinhydrone electrode, or glass electrode in conjuction with a reference electrode, usually a calomel electrode.

EXPERIMENT – 29 Determination pf pH of a Given Solution by pH Metry

INTRODUCTION

Analytical methods that are based on electrode potential measurements are termed potentiometric methods. In this experiment the pH meter will be utilized to measure potentials and pH.

The pH of a substance is a measure of its acidity just as a degree is a measure of temperature. The term pH means hydrogen ion exponent and is defined in terms of hydrogen ion activity "a".

$$pH = -\log_{10} a_{H^+}$$

The following two methods are generally used for the determination of pH of a solution:

1. ***Potentiometric method:*** The pH of a solution can be accurately determined by this method using an electrode that is reversible to H^+ *e.g.*, Hydrogen electrode, Quinhydrone electrode, Glass electrode, Antimony–antimony oxide electrode, etc. Glass electrode in conjunction with a reference electrode (*e.g.* Standard Calomel Electrode, SCE) is most commonly used for the determination of pH of a solution. The cell so formed cam be represented as follows:

   Ag, AgCl(S)

   Cl (0.1 M) | Glass | Test solution || KCl(satd.), Hg_2Cl_2(S) | Hg

   The E.M.F. of the complete cell is given by

   $$E_{cell} = E^o + 0.0591 \text{ pH}$$

   $$pH = \frac{E_{cell} - E^o}{0.0591} \text{ at } 25^oC$$

The Glass Electrode

The glass electrode (Fig. 1) is the most widely used hydrogenion responsive electrode. It works on the principle that when a specially prepared glass membrane is immersed in a solution, a potential is developed which is a linear function of the hydrogen ion concentration of the solution. The glass electrode comprises of a bulb, B, which is filled with 0.1 M HCl solution and into which a Silver–Silver chloride electrode is inserted. As long as the concentration of the internal HCl solution is maintained constant, the potential of the Ag–AgCl electrode inserted into it will remain constant; and also the potential between the HCl solution and the inner surface of the glass bulb will be constant. Hence, the only potential which can vary is that existing between the outer surface of the glass bulb and the test solution. Lithium based glasses are used for the hydrogen-ion responsive glass electrodes.

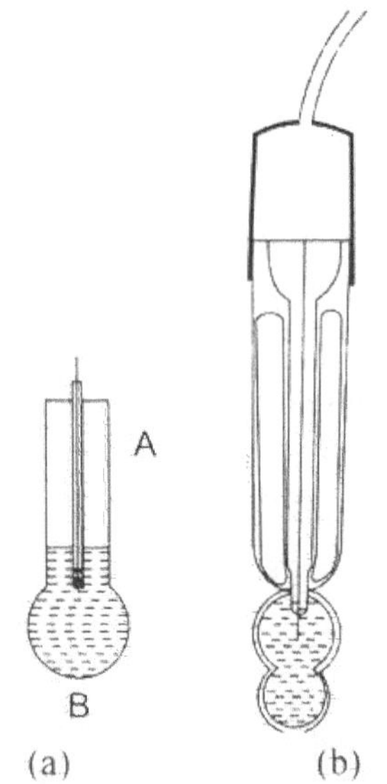

Fig. 1 Glass electrode

pH Meters

Owing to the high resistance of a glass electrode (1 to 100 megohms), a simple potentiometer cannot be employed for measuring the cell E.M.F. and hence the use of special instrumentation is essential. Early pH meters were classified into the following two types: 1. Potentiometric type and 2. Direct reading type.

AIM

To determine the pH of a given solution by pH metry.

APPARATUS

1. pH meter 2. Glass electrode 3. Beaker

CHEMICALS

1. 0.05 M Potassium hydrogen phthalate (pH 4) 2. Unknown pH solution

THEORY

The degree of acidity or alkalinity of a solution is expressed by the pH scale which is a series of numbers between 0 to 14. The tern pH was first coined in the year 1909 by Sorensen and was defined as the negative logarithm of hydrogen ion concentration expressed in molarity:

$$pH = -\log [H^+] \text{ or } -\log C_{H+}$$

However, it is realized that instead of concentration, it is the activity of the ion that determines the e.m.f of a galvanic cell of the type commonly used to measure pH. Hence pH may be defined as the negative logarithm of the hydrogen ion activity:

$$pH = -\log a_{H+} = -\log C_{H+} \times f \pm H^+$$

where f = is the mean ionic activity coefficient.

Although this definition is more consistent. with thermodynamics of the pH electromotive cells, the conventional experimental methods for the determination of pH are incapable of furnishing these thermodynamic quantities. Thus, the term pH is merely a mathematical symbol of convenience devoid of precise thermodynamic validity.

PROCEDURE

(i) Connect the instrument to mains and allow the instrument to warm up for about 10 minutes.

(ii) Prepare three buffer solution; 9.2 pH, 4 pH and 7 pH using standard buffers in distilled water.

(iii) Adjust the temperature dial to the temperature of the soln. and also put the MODE switch in pH position.

(iv) Dip the electrode fully (reference tip also has to be immersed) in 7.00 pH soln. and wait for reading to be stabilised. Adjust the Asymm. Pot. Knob (with screwdriver) to get a reading of 7.00

(v) Now immerse the electrode in either 9.2 pH or 4 pH buffer and adjust the 'SLOPE' knob so that display reads the pH of the buffer and tighten the nut.

(vi) Repeat step (iv) and (v) again, each time electrodes are to be washed in distilled water.

(vii) To check, immerse the electrode in third buffer to verify. (Stirrer may be put ON for better results)

(viii) Now wash the electrode with distilled water. Then insert the electrode in the beaker containing unknown pH solution. Note down the pH reading.

IMPORTANT

For good result, calibration should be done with 7.00 pH buffer and 9.2 pH buffer while working with soln. lying between 7 to 14 pH and 4.00 pH buffer and 7.00 pH buffer should be used for 0-7 pH working.

PREPARATION OF BUFFER FOR VERY ACCURATE MASUREMENT

1. Prepare double distilled water. Keep it closed to avoid CO_2 mixing. Preferably boil it before use.
2. Take a clean beaker (cleaned by distilled water) and place buffer powder or tablet in it. Now add above prepared distilled water so that the total volume is 100 ml. Allow it to dissolve and cover it at the same time to avoid CO_2 mixing. The buffer so prepared will be very accurate.

SOME HINTS FOR STABLE READINGS

Of all the instruments pH - electrode system requires great understanding and care.

We incorporate from time to time latest information gathered by us on the above subject.

If the fluctuation does not stop then try the following:

1. Check the level of KCl. Add freshly prepared KCl each time if possible. The KCl has to flow out of fibre junction freely.
2. See that the tip of the fibre junction dips in the solution.
3. Remove any bubble in the glass bulb.
4. Prick rubber cap for free flow of air. or open the cap and close it for pressure equalisation.

RESULT

The pH of a given solution = ———

EXPERIMENT – 30 Potentiometric Titration of Strong Acid and Strong Base

AIM

To determine the normality of HCl by titrating with NaOH using potentiometer.

APPARATUS

1. pH meter
2. Glass electrode
3. Micro burette
4. Beaker
5. Volumetric flasks
6. Glass rod

CHEMICALS

1. 0.1N Hydrochloric acid (HCl)
2. 0.1N Sodium hydroxide (NaOH)

THEORY

When a solution of acid is titrated with the solution of an alkaline, the change in the pH will be reflected in the change of 'E' (potential). When a small amount of standard alkaline is added to the acid, a little change in the EMF is produced in the beginning. The change in the electrode potential depends upon the fraction of hydrogen ions removed. As an equivalence point reaches the fraction of the hydrogen ions removed by constant volume of standard alkali increases rapidly. There by causing a rapid change in the EMF. Above the equivalence point there is again small change in the EMF by the addition of excess of alkaline. Thus if the EMF of the cell is plotted against the volume of the standard alkali added a curve is obtained.

$$HCl + NaOH \rightarrow NaCl + H_2O$$

The point of intersection in the curve (the point where that curve changes its curvature) gives the equivalence point. It may be noted that the changes in the EMF is much more rapid near the equivalence point than any other region of the titration before and after the equivalence point. This principle applies to all potentiometry titrations, the general form of the titration curve being the same for all. When titration curve does not show a sharp intersection point, the exact location of the point of intersection is rather difficult. The precise method is the differential method where $\Delta E/\Delta V$ values change in E (pH) resulting from the successive additions of the reagent is plotted against the volume of the reagent added. The maximum of the curve (differential curve) so obtained corresponds to the equivalence point of the titration.

$$H_2\,(Pt)\,/\,\text{acid solution}\,//KCl\,(aq.)/\text{calomel electrode.}$$

The EMF of the cell is given by

$$E = E_{cal} - E_{H} = E^{1} + 0.0591\,pH$$

PREPARATION OF REAGENTS

1. ***0.1 N HCl:*** Dilute 0.86 ml of 11.6 N and 36 % concentrated HCl (Mol. Wt. 36.5) to 100 ml with distilled water.
2. ***0.1 N NaOH:*** Dissolve 0.4 gm of NaOH (Mol. Wt. 40) in 100 ml of distilled water.

PROCEDURE

(i) Calibrate the instrument before starting the experiment.

(ii) Take 20 ml of acid solution in a 50 ml beaker and immerse the pH electrode into the solution.

(iii) First carry out the rough titrations by adding 1ml of sodium hydroxide and measure the EMF at each stage.

(iv) After finding the range of end point, then take fresh sample of acid solution in a beaker and carry out the titrations repeatedly by adding 5 ml, 3 ml, 2 ml, 0.5 ml, 0.2 ml and 0.1 ml successively.

(v) After the end point, the volume of alkaline added is increased to 1 or 2 ml. Near the end point smaller additions should be added by micro burette.

(vi) Plot pH values or EMF values against the volume of sodium hydroxide (NaOH) added. Draw a smooth curve, the point of intersection gives the **equivalence point.**

(vii) Plot another graph between ΔE/ΔV values against the titre readings as abscissa. The maximum of the curve represents the equivalence point.

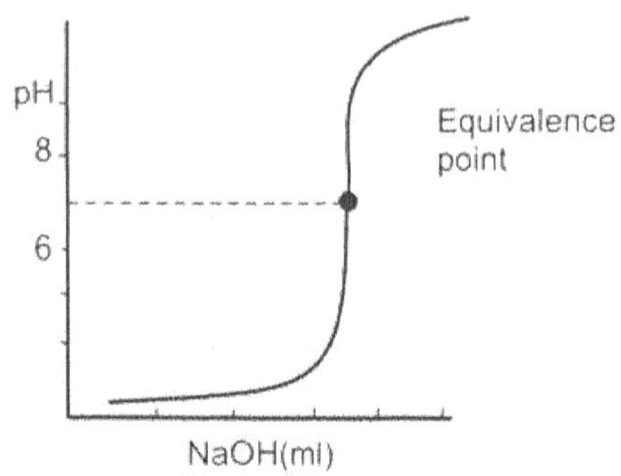

Titration of a strong acid (HCL) with a strong Base (NaOH)

Fig. 1

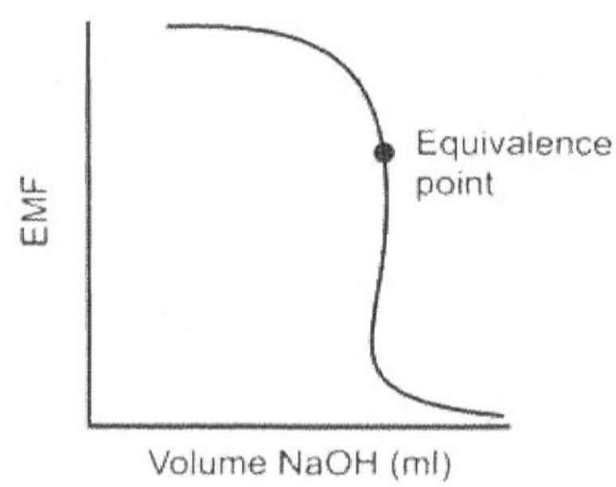

Fig. 2

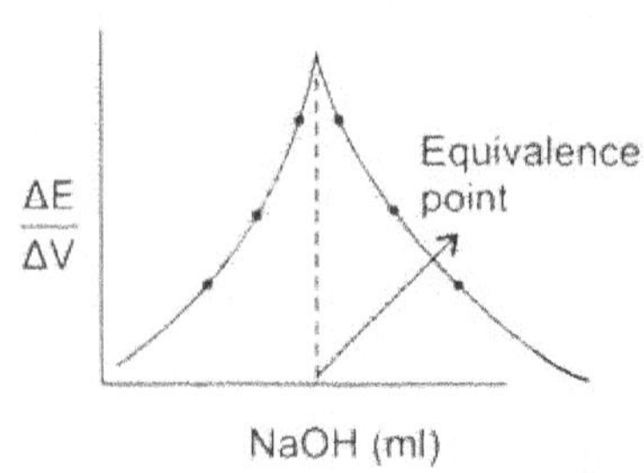

Fig. 3

OBSERVATIONS AND CALCULATIONS

Volume of base added (ml)	pH	mv	$\Delta V\ (V_2 - V_1)$	$\Delta E(E_2 - E_1)$	$\Delta E/\Delta V$

$$\text{Normality of HCl} = \frac{\text{Vol. of NaOH (equivalent point)} \times \text{Normality of NaOH}}{\text{Vol. of HCl taken (20 ml)}}$$

RESULT

The normality of HCl by titrating with NaOH using potentiometer is = __________ N

EXPERIMENT - 31 Potentiometric Titration of Weak Acid Vs Strong Base

AIM

To determine the normality of acetic acid by titrating with sodium Hydroxide using potentiometer

APPARATUS

1. Beaker 2. Conical flask 3. Micro burette 4. pH meter.

CHEMICALS

1. 0.1N Acetic acid 2. 0.1 N Sodium Hydroxide.

THEORY

The EMF of the cell is given by

H_2 (Pt) / acid solution // KCl (aq.)/calomel electrode

$$E = E^1 + 0.0591pH$$

When a solution of an acid is titrated with the solution of an alkaline, the change in the pH will be reflected in the change of 'E' (potential). When a small amount of standard alkaline is added to the acid, a little change in the EMF is produced in the beginning. The change in the electrode potential depends upon the fraction of hydrogen ions removed. As equivalence point reaches the fraction of the hydrogen ions removed by constant volume of standard alkali increases rapidly. There by causing a rapid change in the EMF. Above the equivalence point there is again small change in the EMF by the addition of excess of alkaline. Thus if the EMF of the cell is plotted against the volume of the standard alkali added a curve is obtained.

$$CH_3COOH + NaOH \rightarrow CH_3COONa + H_2O$$

The point of intersection in the curve (the point where that curve changes its curvature) gives the equivalence point. It may be noted that the changes in the EMF is much more rapid near the equivalence point than in any other region of the titration before and after the equivalence point. This principle applies to all potentiometry titrations, the general form of the titration curve being the same for all. When titration curve does not show a sharp intersection point, the exact location of the point of intersection is rather difficult. The precise method is the differential method where $\Delta E/\Delta V$ values change in E (pH) resulting from the successive additions of the reagent(0.2 to 0.4ml) is plotted against the volume of the reagent added. The maximum of the curve (differential curve) so obtained corresponds to the equivalence point of the titration.

PREPARATION OF REAGENTS

1. ***0.1 N Acetic acid:*** Dilute 5.74 ml of 17.4 N and 99.5% acetic acid to 1 litre with distilled water.
2. ***0.1 N NaOH:*** Dissolve 0.4 gm of NaOH (Mol. Wt. 40) in 100 ml of distilled water.

PROCEDURE

(i) Calibrate the instrument before starting the experiment.

(ii) Take 20ml of acid solution in a 50ml beaker and immerse the pH electrode into the solution (V_1).

(iii) First carry out the rough titrations by adding 1ml of sodium hydroxide and measure the EMF at each stage.

(iv) After finding the range of end point, then take fresh sample of acid solution in a beaker and carry out the titrations repeatedly by adding 5ml, 3ml, 2ml, 0.5ml, 0.2ml and 0.1ml successively.

(v) After the end point, the volume of alkaline added is increased to 1 or 2ml. Near the end point smaller additions should be added by micro burette.

(vi) Plot pH values or EMF values against the volume of sodium hydroxide (NaOH) added. Draw a smooth curve, the point of intersection gives the equivalence point.

(vii) Plot another graph between ÄE/ÄV values against the titre readings as abscissa. The maximum of the curve represents the equivalence point.

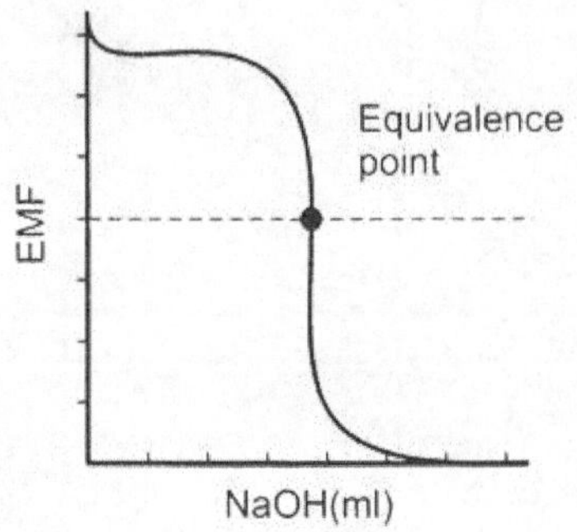

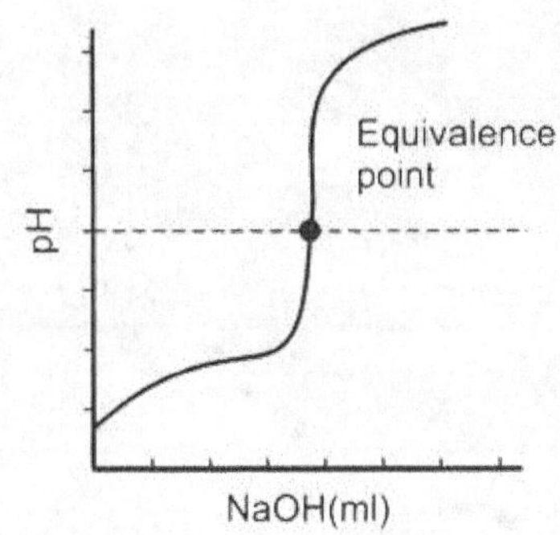

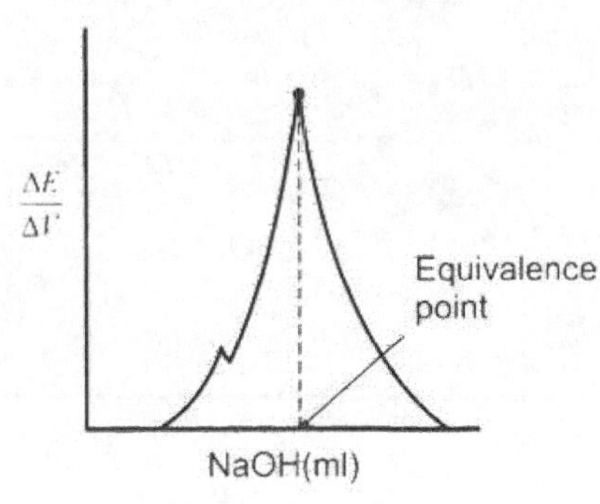

Volume of base added (ml) (V_2)	pH	mv	ΔV ($V_2 - V_1$)	ΔE ($E_2 - E_1$)	$\Delta E/\Delta V$

OBSERVATIONS AND CALCULATIONS

$$\text{Normality of Acetic acid } (CH_3COOH) = \frac{\text{volume of NaOH} \times \text{Normality of NaOH}}{\text{Volume of } CH_3COOH \text{ taken}}$$

RESULT

The normality of CH_3 COOH = __________ N

EXPERIMENT – 32 Determination of Dissociation Constant of Weak Acid

INTRODUCTION

Dissociation constants of weak acids and bases can be determined by measuring the pH of a solution containing known amounts of the acid (base) and its salt with a strong base (acid).

Consider a weak monobasic acid HA which ionises as:

$$HA \rightleftharpoons H^+ + A^-$$

The dissociation constant, K_a, of the acid will be

$$K_a = \frac{a_{H^+} \times a_{A^-}}{a_{HA}} = \frac{C_{H^+} C_{A^-}}{C_{HA}} \cdot \frac{\gamma_{H^+} \cdot \gamma_{A^-}}{\gamma_{HA}} \qquad .....(1)$$

In the presence of highly ionised salt, C_{A-} is equal to the concentration of the salt. a_{H+} can be determined by using a suitable form of hydrogen electrode. γ_{HA} can be taken as unity and γ_A^- can be calculated from Debye-Huckel equation. Thus K_a can be determined.

Dibasic acid ($K_1 > 100\ K_2$). Dissociation of the dibasic acid H_2A takes place in two steps:

$$H_2A \rightleftharpoons H^+ + HA^-$$
$$HA^- \rightleftharpoons H^+ + A^{2-}$$

and

If the two dissociation constants are well separated (i.e., when $K_1 > 100\ K_2$), the firt dissociation is coniplete before the second commences. Thus in the titration two equivalence points, i.e., the inflexion points corresponding to two volume readings in the ratio of 1 : 2 will be obtained.

According to the theory identical with that given for a mono-basic acid pK_1= pH when half an equivalent of alkali corresponding to first equivalence point is added. It can be shown that pK_2=pH when 1½ equivalents of alkali correspording to second equivalence point, have been added, i.e. $[A2^-] = [HA^-]$. More accurately, pK_2 can be obtained by plotting log $[A^{2-}]/[HA^-]$ against pH (ordinate) and extrapolating the linear graph, whence the intercept = pK_g.

For a tribasic acid with groups separated, such as H_3PO_4, the third dissociation constant will be obtained from the pH at 2½ equivalents of alkali added.

AIM

To obtain the dissociation constant of given weak acid.

APPARATUS

1. Potentiometer
2. Glass electrode
3. Beakers
4. Burette
5. Pipette

CHEMICALS

1. 0.1N Acetic acid
2. 0.2 N NaOH

THEORY

When acetic acid is titrated against a base, the H^+ ion concentration in the solution decreases and hence the emf of the cell also decreases. When emf of the cell is plotted against the volume of the base, a curve showing a decrease in emf is obtained. The volume of the base corresponding to a steep decrease in emf gives the titre value. Divide the titre value into two equal parts, the emf of the cell at 1, $\frac{1}{2}$, neutralizations can be noted from the graph. For each emf value thus noted, the corresponding pH of the solution can be calculated using Nernst equation.

or $E = E^0 - \frac{RT}{nF} lnH$ which is the Nernst Equation

at 25^0C and substituting in 2.303 log Q which equals ln Q

we get the more familiar form of the Nernst Equation:

$$E = E^0 - \frac{0.0592}{n} \log H^+$$

Please note that the above equation is only valid at 25°C, you should use the other form of the Nernst equation at any other temperature

$$\mathbf{CH_3COOH + NaOH \rightarrow CH_3COONa + H_2O}$$

During the course in the neutralization, the weak acid and its strong salt coexist in the solution thus making it an acidic buffer

$$CH_3COOH \overset{Ka}{\rightleftharpoons} CH_3COO^- + H^+$$

$$CH_3COONa \rightleftharpoons CH_3COO^- + Na^+$$

K_a is the dissociation constant of the acetic acid. The pH of an acidic buffer is given by the Henderson equation.

$$pH = p^{Ka} + \log \frac{[CH_3COONa]}{[CH_3COOH]}$$

$$pH = p^{Ka} + \log \frac{[salt]}{[Acid]}$$

Determining potentiometrically, the pH of the solution at ¼, ½, neutralizations, the pK_a values for these can be computed. From each pK_a, the dissociation constant of the acid can be found.

$$p^{Ka} = -\log ka$$

PROCEDURE

(i) Place 20 ml of the acetic acid solution in a small beaker and find the emf by inserting the glass electrode.

(ii) Calculate the standard emf, E_0 of the cell.

(iii) Add 0.1 ml of NaOH from the burette to the acid and stir the solution. Determine the emf untill to get the end point.

(iv) Plot the emf against NaOH added.

(v) Divided it into two equal portions and mark in the graph. the volumes of NaOH required for ¼, ½ neutralizations.

(vi) Note down the emf corresponding to each volume and calculate the pH values from the Nernst equation.

(vii) Obtain for eah pH. the p^{Ka} using the Henderson equation and calculate k_a for each trial.

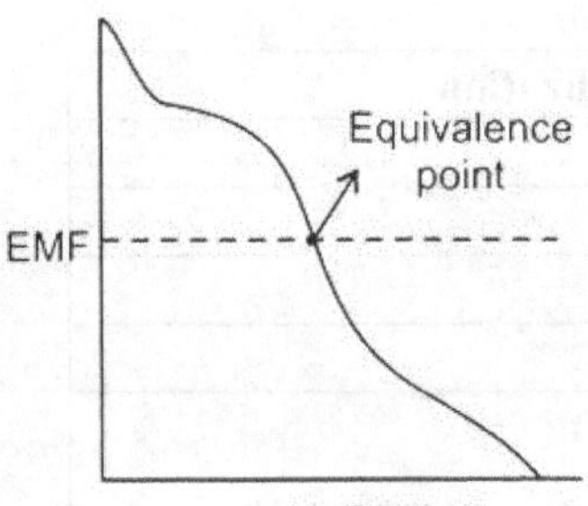

Fig. 1

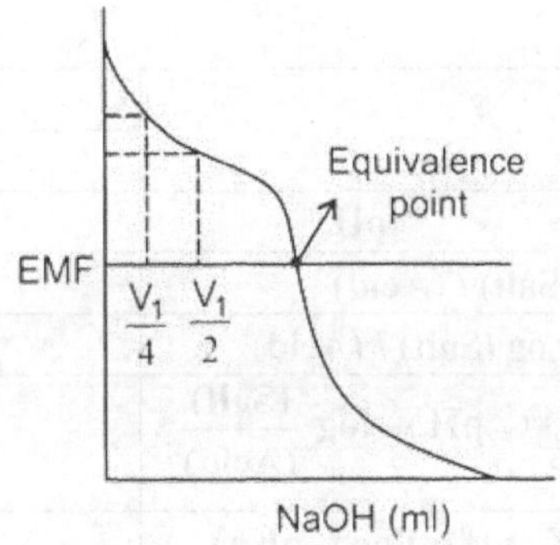

Fig. 2

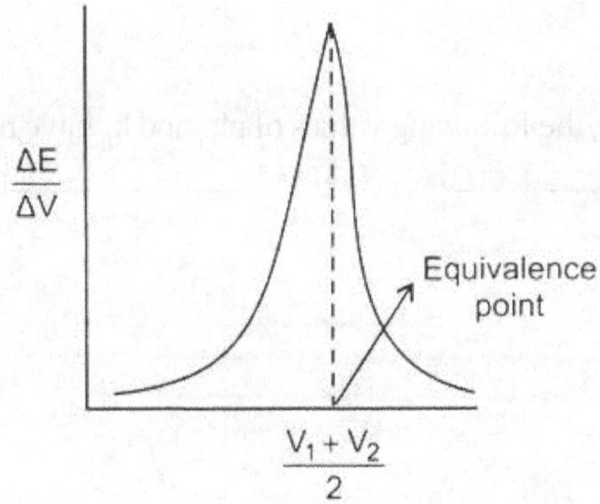

Fig. 3

OBSERVATIONS AND CALCULATIONS

Volume of base added (ml)	pH	mv	ΔV ($V_2 - V_1$)	ΔE($E_2 - E_1$)	$\Delta E/\Delta V$

Stage of Neutralization	EMF (volts)	pH
1/4		
1/2		

	Stage of Neutralization	
	1/4	1/2
pH		
(Salt) / (Acid)		
Log (Salt) / (Acid)		
$p^{ka} = pH - \log \frac{(Salt)}{(Acid)}$		
K_a = Antilog (–pKa)		

RESULT

At various stages of neutralisation, the following values of pk_a and k_a have been obtained for the dissociation of acetic acid is gas ka_1 = ____________ ka_2 = ____________

CHEMICAL KINETICS

INTRODUCTION

It is a well known fact that the chemical reactions occur with widely different but definite rates. The rate of a reaction, however, depends upon the various experimental conditions such as the concentration of the reactants, temperature, presence of catalyst etc.

Reaction rates: The rate of any reaction is the rate of change of concentration of any one of its reactants or reaction products with time. Since the rate of a reaction depends upon the concentrations of the reactants, decreases as the concentrations decrease, it can not be obtained simply dividing the change in concentration by the time taken for the change.

If dc is the decrease in concentration of one of the reactants in a vanishingly small time interval, dt, the rate at that instant is given by

$$\text{Rate of reaction} = -dc/dt.$$

The minus sign implies that the concentration of the reactant decreases as the reaction proceeds. More generally, the rate is represented by dx/dt, where dx is the number of moles of either reactant consumed or product formed in time interval, dt.

Order of reaction: It may be defined as the number of atoms or molecules whose concentration determine the rate of the reaction. The reactions are classified as those of first, second or third order according to the number of atoms or molecules under going the change is one, two or three respectively. The term order of a reaction may further be understood from the fact that if the rate of reaction is proportional to certain power, a, of reacting species the order of reaction with respect to that reactant is said to be a; the order of overall reaction will be the sum of the orders with respect to each reactant involved.

First order reaction: By definition the rate is proportional to the first power of concentration of the reactant. *i.e.*

$$-\frac{dc}{dt} \propto C, \text{ or } -\frac{dc}{dt} = kC. \qquad .....(1)$$

Where k is called the veocity constant, rate constant or specific reaction rate, which is equal to the rate of reaction when the concentration is unity.

Suppose we start with a g moles/litre of the reacting substance and x g moles of it transform into the products after a time, t. The rate of the reaction at time, t, will be given by

$$-\frac{dc}{dt} = \frac{d(a-x)}{dt} = \frac{dx}{dt} = k(a-x) \qquad .....(2)$$

Integrating eq. (2) and using the fact that when $t = 0$, $x = 0$, we have

$$k = \frac{1}{t} \ln \frac{a}{a-x} = \frac{2.303}{t} \log \frac{a}{a-x} \qquad .....(3)$$

If x_1 and x_2 are the moles of the reactant changed into the products at time t_1 and t_2, then eq. (3) takes the form

$$k = \frac{2.303}{t_2 - t_1} \log \frac{a-x_1}{a-x_2} \qquad ....(4)$$

Equations (3) and (4) are the the expressions for the rate constant of first order reaction. Eq. (4) is of particular use when the initial concentration, a, is not known.

Second order reaction: In a second order reaction the rate is proportional to the second power of concentration of the reactant. A second order reaction can be represented in either of the two ways:

$$2A \longrightarrow \text{Produts} \quad .....(6)$$

$$A + B \longrightarrow \text{Products} \quad .....(7)$$

The velocity of the reaction is given by

$$-\frac{dc}{dt} = kC_A^2 \qquad \text{For reaction (6)}$$

and

$$-\frac{dc}{dt} = kC_A C_B \qquad \text{For reaction (7)}$$

Suppose we start with equal concentrations of A and B, say a moles/litre, or simply, a moles/litre in case the reaction is represented by eq. (6). The velocity of the reaction at any instant t will be

$$\frac{dx}{dt} = k\ (a - x)^2 \quad .....(8)$$

where x is the number of moles of each reactant changed into the products.

Remembering that $x = 0$ when $t = 0$, integration of eq. (8) gives,

$$k = \frac{1}{t} \cdot \frac{x}{a(a-x)} = \frac{1}{t}\left(\frac{1}{a-x} - \frac{1}{a}\right) \quad .....(9)$$

When the initial concentrations of the two reactants in the reaction represented by eq. (7) are a and b moles/litre and x moles of each transform into products any time t, the rate equation will then be

$$\frac{dx}{dt} = k(a - x)\,(b - x) \quad .....(10)$$

Again integrating of eq. (10), with the limits $x = 0$ when t = 0, gives

$$k = \frac{2.303}{t\,(a-b)} \log \frac{b\,(a-x)}{a\,(b-x)} \quad .....(11)$$

Expressions (9) and (11) and the two forms of rate equation for the second order reaction.

Third order reaction: A third order reaction may be represented as

$$3\,A \longrightarrow \text{Products}$$

If a moles/litre is the initial concentration of the reaction and x moles/litre change into products at any time, t, then the rate is given by

$$\frac{dx}{dt} = k\,(a - x)^3$$

On carrying out the integration of the above expression with the limits x = 0 when $t = 0$, we get

$$k = \frac{1}{2t} \cdot \frac{x\,(2a - x)}{a^2\,(a - x)^2} = \frac{1}{2t}\left\{\frac{1}{(a-x)^3} - \frac{1}{a^2}\right\} \quad .....(12)$$

Time required to complete the same fraction of the change, say half ($x = 0.5a$), may be obtained by putting $x = 0.5a$ in (12)

$$t_{1/2} = \frac{1}{2k} \cdot \frac{\frac{a}{2}\left(2a - \frac{a}{2}\right)}{a^2\left(a - \frac{a}{2}\right)^2} = \frac{3}{2a^2k}$$

Thus time taken for any definite change, say half change, is inversly proportional to the square of the initial concentration.

The temperature coefficient (Energy of activation): Increase in temperature almost in all cases increases the velocity of a reaction to a marked extent. It has been seen that in general the specific rate is approximately doubled or tripled for an increase in temperature by 10°C. The ratio of specific rates at temperatures separated by 10°C, usually 25° and 35°C, is called the temperature coefficient of the teacrion rate. That is

Temperature coefficient = $k_{t+10} \approx 2$ to 3 (18)

where k_t is the specific rate at t°C and k_{t+10} at 10°C higher.

The above method is approximate for indicating the influence of temperature as the coefficient decreases with increasing temperature. The variation of veiocity constant is best expressed in the form of an exponential equation, known as Arrhenius equation:

$$k = Ae^{-E/RT} \qquad .....(19)$$

where E is the energy of activation (cal/mole) and A is a constant. Taking logarithm of the above equation, we have

$$2.303 \log k = -\frac{E}{RT} + \text{constant}$$

If k_1 and k_2 are specific rates at two temperatures T_1 and T_2K.

then

$$\log\frac{k_2}{k_1} = -\frac{E}{2.303R}\left[\frac{1}{T_2} - \frac{1}{T_1}\right] \qquad .....(20)$$

$$= -\frac{E}{4.576}\left[\frac{1}{T_2} - \frac{1}{T_1}\right]$$

Thus the measurement of specific rates of a reaction at two temperatures enables E to be calculated.

If on the other hand the values of specific rates at a series of temperatures are known, energy of activation, E, can be obtained by plotting log k values (ordinates) against 1/T. The graph will be a straight line with slope equal to –E/2.303 R.

EXPERIMENT - 33 Determination of Rate Constant and Activation Energy of Given Ester Catalysed by an Acid

AIM

To determine the rate constant of hydrolysis of an ester such as methyl acetate catalysed by an acid (0.5 M HCl) and also determine the energy of activation.

APPARATUS

1. Burette
2. Conical flask
3. Iodination flask
4. Plastic tray for thermostat
5. Pipette
6. Stop watch
7. Water bath

CHEMICALS

1. Methyl acetate
2. 0.5M HCl
3. 0.1M NaOH
4. Phenolphthalein indicator

THEORY

Methyl acetate is hydrolyzed to give methyl alcohol and acetic acid as shown below:

$$CH_3COOCH_3 + H_2O \xrightarrow{H^+} CH_3COOH + CH_3OH$$

The reaction is catalyzed by hydrogen ions. In a dilute aqueous solution of the ester concentration of water being excess, is very high and practically remains constant during the reaction; concentration of H^+ which catalyze the reaction also remains constant. Thus the rate of reaction is dependent only on the conc. of the ester i.e.,

$$dx/dt = k[CH_3COOCH_3]$$

Following the kinetic equation of first order.

$$k = 2.303/t \quad \log a/a-x$$

Since acetic acid is produced as a result of hydrolysis, the kinetics of reaction can be followed by withdrawing affixed volume of the reaction mixture from time to time and titrating a standard alkali. The titre value is equivalent to the some of the acid used as the catalyst, which remains constant throughout, and of the acetic acid produced during the reaction. The difference of titre values at any time after the commencement of the reaction and that at the commencement gives the acetic acid formed and hence the amount of methyl acetate hydrolyzed at that instant.

PREPARATION OF REAGENTS

1. ***0.5M HCl:*** Dilute 4.3 ml of 11.6 N and 36 % concentrated HCl (Mol. Wt. 36.5) to 100 ml with distilled water.
2. ***0.1M NaOH:*** Dissolve 0.4 gms of NaOH (Mol. Wt. 40) in 100 ml of distilled water.
3. ***Phenolphthalein Indicator:*** Dissolve 1gm of phenolphthalein in 100 ml of distilled ethanol and add 100 ml of distilled water with constant stirring, filter, if necessary.

PROCEDURE

(i) Take 100 ml 0.5M HCl in a dry 250 ml conical flask and clamp it in the thermostat at 30^0 C. Suspend also a stoppered tube in thermostat and place in it about 15 ml methyl acetate.

(ii) Fill a burette with 0.1M NaOH solution, previously rinsing it with the same solution. Take 3-4 conical flasks containing about 250 ml of ice cold water with small pieces of clean ice.

(iii) When temperature equilibrium has been reached, pipette about 5 ml of ester in to the conical flask of the acid. Shake well and immediately pipette 5 ml sample of the reaction mixture in to a conical flask containing ice cold water so as to arrest the reaction; record the time to the nearest 10 seconds when the pipette has been half discharged in to the flask. Titrate this solution as rapidly has possible with 0.1M NaOH using phenolphthalein indicator. The titre value gives the amount of HCl in the sample at the start of the reaction.

(iv) Make similar titrations of further 5 ml samples at successive intervals of 10, 20, 30, 40, 60, 90, and 120 min.

(v) Allow the remaining reaction mixture to stand for 48 hours so that reaction goes to completion. Titrate 5 ml sample with alkali as before. If such as a long time is not available place the reaction vessel containing the remaining reaction with its cork closed loosely, in water bath at about 50^0 C for at least 1 hr to complete the reaction. Then after cooling the flask to room temperature, titrate 5 ml of sample as before. Much time can be saved if 25 ml of the reaction mixture is placed in well stoppered small flask at about 50^0 C immediately after taking the first reading

(vi) Repeat the experiment at different temperatures, say 40^0 C, for the calculation of energy of activation

PRECAUTIONS

(a) Temperature of reaction mixture should be constant to within $\pm 0.05^0$ C throughout the course of the reaction.

(b) The reaction must be stopped before carrying out the titration of the sample by adding it to ice-cold water.

(c) Titration should be made as rapidly as possible.

OBSERVATIONS AND CALCULATIONS

Time in Minutes	Titre Reading (ml)	x ml	a-x ml	log (a-x)	k

As each molecule of ester is hydrolyzed 1 molecule of acetic acid is produced. The acid concentration, therefore, increases as the reaction proceeds. This increase in acid conc. is a direct measure of amount of ester hydrolyzed i.e., x. If T_0 & T_∞ are titre readings, at times zero, t and infinitive (after 48 hrs when the reaction is complete), then initial conc., a, of the ester is proportional to the $T_\infty - T_0$ and that at time t, i.e. $(a - x)$ to $T_\infty - T_t$. The first order rate equation is

$$a = T_\infty - T_0$$

$$a - x = T_\infty - T_t$$

$$K = 2.303/t \quad \text{Log} \frac{(T_\infty - T_0)}{(T_\infty - T_t)}$$

$$t = 2.303/k \; \{\log(T_\infty - T_0) - \log(T_\infty - T_t)\}$$

OR

Calculate the values of k by making substitution of titre values in the above expression and take the average.

Or plot t (abscissa) verses $\log(T_\infty - T_t)$ and obtain the value of k from the slope, $-k/2.303$, of the straight line so obtained.

From the values of reaction rate constant at two temperatures calculate the energy of activation for the reaction using the equation:

$$\log \frac{k_2}{k_1} = -\frac{E}{2.303R}\left[\frac{1}{T_2} - \frac{1}{T_1}\right]$$

$$\log \frac{k_2}{k_1} = -\frac{E}{4.567}\left[\frac{1}{T_2} - \frac{1}{T_1}\right]$$

$$E = \frac{T_1 T_2}{T_2 - T_1} \times 4.567 \times \log \frac{k_2}{k_1}$$

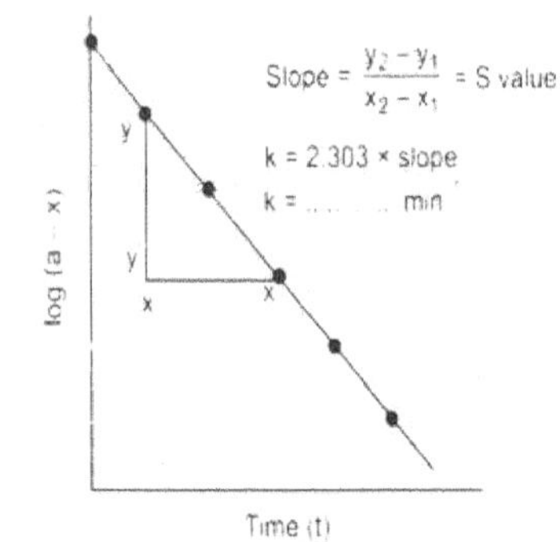

RESULT

The rate constant of the reaction (k) = ———

The energy of activation (E_a) = ———

EXPERIMENT – 34 Study the Kinetics of the Reaction between Potassium Persulphate and Potassium Iodide

AIM

To determine the rate constant and order of the reaction . Also study the influence of ionic strength on the rate constant.

APPARATUS

1. Conical flasks
2. Pipette
3. Plastic tray for thermostat
4. Burette
5. Iodination flask

CHEMICALS

1. 0.05 M KI solution
2. 0.05 M $K_2S_2O_8$ solution
3. 0.004 M $Na_2S_2O_3$ solution
4. 0.4 M KCl solution
5. 1% Starch solution

THEORY

The reaction between $K_2S_2O_8$ and KI :-

$$K_2S_2O_8 + 2KI = 2K_2SO_4 + I_2$$

(or)

$$S_2O_8^{-2} + 2I^- = 2\ SO_4^{-2} + I_2$$

Is found to be of second order. The first stage of the reaction is probably a bi molecular slow stage

$$S_2O_8^{2-} + I^- = [S_2O_8I]^{-3} \quad \text{(Slow step)}$$

Followed by the rapid de composition of triply charged anion

$$[S_2O_8I]^{3-} + I^- = I_2 + 2SO_4^{2-} \quad \text{(Fast step)}$$

Since the reaction is followed with the liberation of iodine, the progress of the reaction is followed by titrating the liberated iodine with a standard $Na_2S_2O_8$ solution using starch solution as the indicator. The titre value at any time (t) will be proportional to the amount of I_2 liberated i.e., to the amount of KI oxidized or the amount of $K_2S_2O_8$ used up in oxidation; hence it gives the value of (**x**) at that time

The influence of ionic strength on K

The change in the ionic strength of the solution influences the specific reaction rate depending upon the signs of the charge on the reacting species. A relationship between the ionic strength **μ** and the rate constant (**k**) is given by **Bronsted-Bjerrum equation**

$$\log k = \log k_0 + 1.02\ Z_A Z_B \sqrt{\mu}$$

where K_0 is the specific rate in infinitely dilute solution; Z_A and Z_B are the charges on the reacting species **A** and **B** (ions)

$$\mu = \frac{1}{2} \Sigma C_i Z_i^2$$

C_i = concentration of particular ions in moles/l

Z_i = valence of same ion

Σ = summation of different kinds of ions in solutions

If Z_A and Z_B are both positive or both negative, **K** will increase with **μ**; it will decrease if only one of Z values is negative. It will remain unaffected if one of reacting species is uncharged. In the present case of $Z_A Z_B = 2$. So, **log K** increases with" $\sqrt{\mu}$.

PREPARATION OF REAGENTS

1. *0.05 M KI solution:* Dissolve 8.3 gms of Iodate free KI in distilled water and make up the volume to 1 litre.
2. *0.05 M $K_2S_2O_8$ solution:* Dissolve 13.5 gm of $K_2S_2O_8$ (Mol. Wt. 270 & Eq. wt. 135) in 1 litre of distilled water.
3. *0.004 M $Na_2S_2O_3$ solution:* Dissolve 0.992 gm of $Na_2S_2O_3 . 5H_2O$ (Mol. Wt. 240.80 & Eq. wt.240.80) in 1 litre of distilled water.
4. *0.4 M KCl solution :* Dissolve 29.824 gm of KCl (Mol. Wt. 74.55 & Eq. wt.74.55) in 1 litre of distilled water.
5. *1% Starch solution:* Make a paste of 1 gm of soluble starch in 5-7 ml of water and pour the paste with constant stirring in 100 ml of boiling water. Allow the solution to cool to room temperature.

PROCEDURE

(i) Place 20 ml of 0.05M $K_2S_2O_8$ solution in a 150 ml conical flask and add to it 60 ml water. Suspend the flask in a thermostat at 25°C. Take about 100 ml solution in an another flask clamped in the thermostat.

(ii) When the solution have acquired the equilibrium temperature, pipette 20 ml KI solution into $K_2S_2O_8$ solution. Start the stop watch when the pipette is half discharged.

(iii) Mix well and withdraw 10 ml samples at intervals of 5. 10. 15. 20. 25 30. 40 minutes and freeze them by running in ice cold water. Titrate each sample against standard $Na_2S_2O_3$ solution as rapidly as possible using starch solution as the indicator.

(iv) Repeat the experiment with the addition of KCl solution to change the ionic strength. Add 20, 40 and 50 ml KCl solution and the appropriate volume of water to make the overall volume of the reaction mixture to 100 ml.

OBSERVATIONS AND CALCULATIONS

The initial concentration of KI = initial concentration of $K_2S_2O_8$ = 0.01 M.

If 10 ml of the reaction mixture is titrated, then the initial concentration of $K_2S_2O_8$ or KI in terms of titre value of 0.004 M $Na_2S_2O_3$ will be

$$V = \frac{10 \times 0.01}{0.004} = 25\text{ml} \equiv a$$

As the concentrations of both the reactants are the same, therefore we have

$$k = \frac{1}{t} \cdot \frac{x}{a(a - x)}$$

If T_t is the titre reading at any time t, then $x = T_t$ and $a - x \equiv 25 - T_t$

Hence $$k = \frac{1}{at} \cdot \frac{T_t}{25 - T_t}$$

Plot a graph between $\frac{x}{a-x}$ i.e.,

$\frac{T_t}{25 - T_t}$ and t (abscissa). The graph will be a straight line with slope a k, i.e., 0.01 k. Hence reduce K

Plot log k against $\sqrt{\mu}$ (abscissa). The graph will be a straight line with a slope 2.04

GRAPH

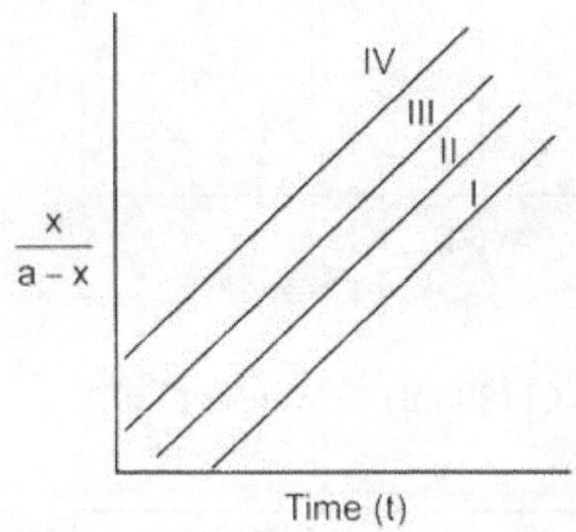

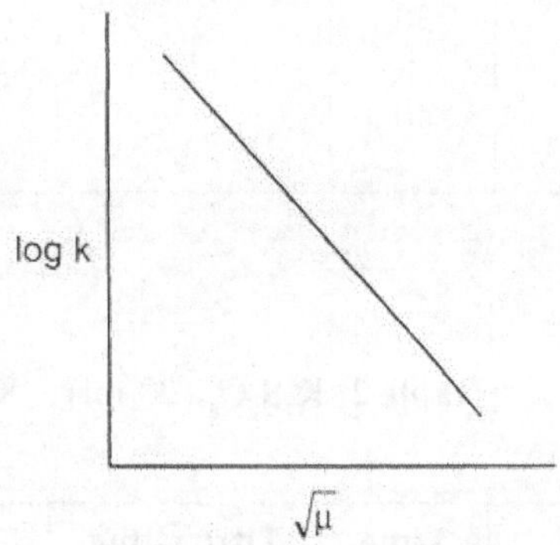

Four k values like

$k_1 = ...$ $k_2 = ...$ $k_3 = ...$ $k_4 = ...$

Four log k values like

$\log k_1 = ...$ $\log k_2 = ...$ $\log k_3 = ...$ $\log k_4 = ...$

Four μ values

$$\mu = \frac{1}{2} \sum c_i z_i^2$$

$$\mu = \frac{1}{2}(0 + 0.08 + 0.16 + 0.20)$$

$$\mu = 0.22$$

Table 1 $K_2S_2O_8$ (20 ml) + KI (20 ml) + KCl (0 ml) + Water (60 ml).

S.No	Time (min)	Titre value (ml)	a(a−x)	x/a(a−x)	$k_1 = \frac{1}{t}\frac{x}{a(a-x)}$
1	5				
2	10				
3	15				
4	20				
5	25				
6	30				
7	40				

Avg K_1 = ———

Table 2 $K_2S_2O_8$ (20 ml) + KI (20 ml) + KCl (20 ml) + Water (40 ml).

S.No	Time (min)	Titre value (ml)	a(a−x)	x/a(a−x)	$k_2 = \frac{1}{t}\frac{x}{a(a-x)}$
1	5				
2	10				
3	15				
4	20				
5	25				
6	30				
7	40				

Avg K_2 = ———

Table 3 $K_2S_2O_8$ (20 ml) + KI (20 ml) + KCl (40 ml) + Water (20 ml).

S.No	Time (min)	Titre value (ml)	a(a–x)	x/a(a–x)	$k_3 = \frac{1}{t}\frac{x}{a(a-x)}$
1	5				
2	10				
3	15				
4	20				
5	25				
6	30				
7	40				

Avg K_3 = ———

Table 4 $K_2S_2O_8$ (20 ml) + KI (20 ml) + KCl (50 ml) + Water (10 ml).

S.No	Time (min)	Titre value (ml)	a(a–x)	x/a(a–x)	$k_4 = \frac{1}{t}\frac{x}{a(a-x)}$
1	5				
2	10				
3	15				
4	20				
5	25				
6	30				
7	40				

Avg K_4 = ———

Result

The rate constant of the reaction (k) = ———

The influence of ionic strength (μ) = ———

EXPERIMENT – 35 Determination of Surface Tension

INTRODUCTION

The property of a liquid that has no counter part in solids or gases is the surface tension. It is the force that tends to minimize the surface area of a drop of liquid or in other words it is the force in dynes acting at right angles to the surface of a liquid or solution. Energy is required to expand the surface of a liquid *and surface tension measures the amount of energy required to increase the surface of a liquid by unit area*. The surface tension of a liquid decreases as the temperature increases. An increase in temperature decreases the intermolecular attraction and therefore less work is required to bring a molecule from the interior of a liquid to the surface. It would be expected then that those liquids with larger intermolecular forces would have greater surface tension. Table given below lists the surface tension of several liquids at 20 °C.

Liquid	Surface tension (dynes/cm)	Liquid	Surface tension (dynes/cm)
Water	72.8	Ethylene glycol	47.7
Benzene	28.9	Glycerol	63.4
Toluene	28.4	Carbon tetrachloride	27.0
Acetone	23.7	Ethyl iodide	29.9
Methyl alcohol	22.6	Ethyl bromide	24.2
Ethyl alcohol	22.3	Nitrobenzene	41.8

AIM

To determine the effect of soap and detergent on the surface tension of water using a Stalagmometer

APPARATUS

1. Stalagmometer
2. Volumetric flask
3. Small rubber tubing
4. Screw type pinch cock

CHEMICALS

1. Detergent solution
2. Soap solution

THEORY

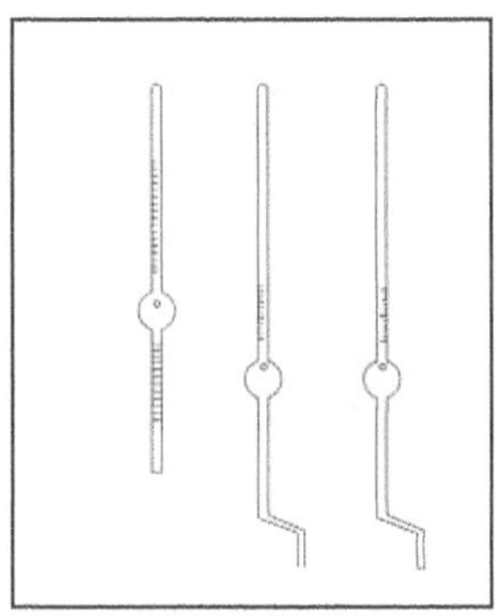

The size of the drop falling off from the end of a tube depends on the surface tension of the liquid and the size of the capillary end, which provides the line of attachment for the drop. Thus when a liquid is allowed to flow through a capillary tube, a drop will increase in size to a certain point and then fall off. The total surface tension supporting the drop is $2\pi r\gamma$ where r is the radius of the outer circumference of the dropping end of the capillary tube. It is along this line that liquid, glass and air meet and the force acting along the circumference, hence $W = 2\pi r\gamma$, where W is the weight of the drop and $2\pi r$ is the circumference of the external wall of the capillary tube.

The surface tension of the liquid can therefore be determined from the weight of a single drop and the external radius of the dropping tube.

If we have two liquids such that

$$W_1 = 2\pi r\gamma_1 \text{ and } W_2 = 2\pi r\gamma_2 \text{ then } W_1/W_2 = \gamma_1/\gamma_2 \quad .....(i)$$

Thus the drop weight method can be employed for comparing the surface tension of two different liquids. However, it is easier to count the number of drops formed by equal volumes of two liquids than finding the weight of a single drop. For two different liquids, the weights of equal volumes are proportional to their densities. Let N_1 and N_2 be the number of drops of liquid produced from the same volume V of the two liquids.

Volume of a single drop = V/N_1

Weight of a single drop = $(V/N_1)d_1$ or $2\pi r\gamma_1 = (V/N_1)d_1$

Similarly for liquid 2

$$2\pi r\gamma_2 = (V/N_2)d_2$$

where d_1 and d_2 are the densities of the respective liquids. Dividing these two equations we get

$$\gamma_1/\gamma_2 = (N_2.d_1)/(N_1.d_2) \quad .....(ii)$$

PROCEDURE

(i) Take a clean and dry Stalagmometer

(ii) Attach rubber tubing with a screw pinch cock to the upper end of the stalagmometer to control the flow of the liquid and clamp it to a stand.

(iii) Make marks above and below the bulb with a marker pen if it is not marked.

(iv) With the help of the rubber tubing, suck distilled water into the stalagmometer till the level rises above the upper mark.

(v) Now allow the liquid to flow through the capillary (control the flow rate using the screw pinch cock) and count the number of drops formed (N_1) till the liquid level reaches the mark below the bulb

(vi) Dilute the given stock soap solution (1%) to appropriate dilutions as given below using a burette and determine the number of drops for each of these concentrations. Tabulate the data as follows

(vii) Repeat the above procedure for 0.5, 0.1 and 0.05 % of detergent solution also and tabulate it as above.

No	Volume of water + Volume of soap/detergent solution (ml)	Concentration of soap/detergent solution	Number of drops			Surface tension (dynes/cm)
			Exp. 1	Exp. 2	Average (N)	
1	50 + 0	Pure water				
2	25 + 25	0.5%				
3	45 + 5	0.1%				
4	47.5 + 2.5	0.05%				

CALCULATIONS

Surface tension of water at 25°C = 72 dynes/cm

r_1 = Surface tension of the reference liquid (i.e., water)

r_2 = Surface tension of the test liquid

$$\frac{r_1}{r_2} = \frac{N_2 . d_1}{N_1 d_2}$$

RESULT

The surface tension of the given liquid = ——— dynes/cm

EXPERIMENT – 36

Determination of the Viscosity of the Given Liquid

INTRODUCTION

When a liquid flows, it has an internal resistance to flow, an internal friction, which is called its viscosity. Viscosity is a characteristic property of all liquids. Honey flows less readily than water because honey has a higher viscosity than water (honey is more viscous than water).

The flow of liquids can be described quantitatively, relatively easily if that flow is laminar, or smooth. Laminar flow is considered to occur as a series of thin plates of liquid sliding past each other smoothly at different velocities. When this is not a reasonable description of the flow, as it would not be for a mountain stream bubbling over rocks, the flow described as non-laminar or turbulent flow.

Liquids whose molecules have strong intermolecular forces have greater viscosities than those liquids that have weaker intermolecular forces. The table below lists the viscosities of some of the common liquids.

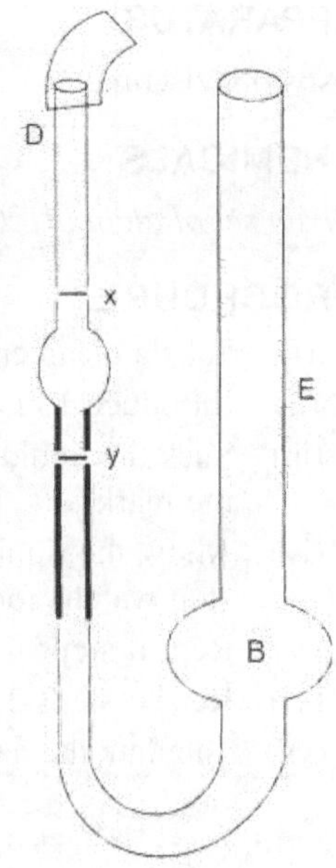

The Ostwald Viscometer

Table 1 Viscosities of some liquids at 20 °C.

Liquid	Viscosity (millipoise)	Liquid	Viscosity (millipoise)
Acetic acid	12.22	Ethyl acetate	4.55
Acetone	3.31	Ethyl alcohol	12.00
Benzene	6.52	Methyl alcohol	5.97
n-Butyl alcohol	29.48	Water (20°C)	10.02
Carbon tetrachloride	9.69	Water (25°C)	8.90
Chloroform	5.80	Water (30°C)	7.975
Glycerol	14900.00		

For any given capillary, r and l are constants. If V is fixed $\eta = k.p.t$ where k includes all constant terms. This equation leads to an easy comparison between the viscosities of different liquids

$$\eta_1 / \eta_2 = k.p_1.t_1 / k.p_2.t_2 = p_1 t_1 / p_2 t_2 \quad .....(i)$$

In the laboratory, the viscometer commonly used for comparing viscosities of liquids is the Ostwald viscometer. In its operation different liquids are taken in exactly the same volume. This is essential so that the height of liquid columns creating the pressure heads are equal and then the pressure heads will be directly proportional to the densities of the respective liquids. The liquids are sucked up into the upper bulb A. Initially enough liquid volume is taken in bulb B such that when the liquid stands above the mark x in bulb A, a little is still left in the bend and the bulb B. The time of flow (t) is measured for each liquid for its flow from x to y. The driving pressure p at all stages of the flow of the liquid is given by h × d × g where g acceleration due to gravity; d = density of liquids. h = difference in the height of the liquids in the upper and lower bulbs (the same for all liquids between the same points x and y)

Equation (iv) can be simplified into

$$\eta_1/\eta_2 = h.d_1.g.t_1/h.d_2.g.t_2 = d_1t_1/d_2t_2 \quad .....(ii)$$

AIM

Determination of the viscosity of given liquid.

APPARATUS

Ostwald Viscometer Wrist watch (timer)

CHEMICALS

Given set of alcohols: (butanol, propanol and ethanol)

PROCEDURE

(i) Take a clean and dry Ostwald Viscometer and set it vertically on a stand.

(ii) Introduce 10 ml of water through E (refer figure above) into the larger bulb B.

(iii) Suck the liquid up into the bulb A through a rubber tubing attached to the end D to a level above the mark x.

(iv) Allow the liquid to flow freely through the capillary and note the time t_1 (use a wrist watch or stop watch) for the liquid to flow from x to y.

(v) Repeat steps 3 and 4 two more times and get an average value of t_1

(vi) Repeat steps 2 to 5 to obtain the average time for the liquids to flow from x to y.

(vii) Tabulate the data as follows

Liquid (density)	Time flow (in sec)		Time (average)	Viscosity (milliopoise) millipoise
	Exp. 1	Exp. 2		
Water (1.0)				
Butanol (0.81)				
Propanol (0.786)				
Ethanol (0.785)				

Use equation (ii) to determine the viscosities of different liquids

Calculations

$$\frac{\eta_1}{\eta_2} = \frac{d_1t_1}{d_2t_2}$$

(t = time of flow from *x* to 7)

RESULT

The viscosity of the given liquid = ——

EXPERIMENT – 37 Determination of Viscosity of Lubricating Oil at Different Temperatures by Redwood Viscometer

INTRODUCTION

Viscosity index

The viscosity of an oil decreases with increase of temperature as result of decrease in intermolecular attraction due to expansion. Hence it is always necessary to state the temperature at which the viscosity was determined.

In many applications lubricating oil will have to function in machinery over a considerably wide range of operating temperatures, if this is due to seasonal variation in atmospheric temperature.

Adjustments can be affected by selecting different oils of appropriate viscosity for different seasons. However, in case of internal combustion engines, aeroplanes etc., the lubricant used must function both at low starting temperatures as well as at very high operating temperatures. Since the viscosity of lubricating oils decreases with temperature, it is impossible to select an oil having same viscosity over such a wide range of operating temperatures. However, one can select an oil whose variation in viscosity with temperature is minimum. This variation can either be indicated by viscosity-temperature curves or by means of the viscosity index. Viscosity index is the numerical expression of the average slope of the viscosity-temperature curve of a lubricating oil between 100°F to 210°F. The oil under examination is compared with two standard oils having the same viscosity at 210°F as the oil under test. Oils of the Pennsylvanian type crude thin down the least with increase of temperature: whereas oils of Gulf coast origin thin down the most as the temperature is increased. Hence the viscosity index of Pennsylvanian oil is taken as 100 and that of the Gulf oil as zero. Then the viscosity of the oil under investigation is deduced as follows.

$$\text{Viscosity index of the oil under test} = \frac{V_L - V_X}{V_L - V_H} \times 100$$

where

V_L = Viscosity at 100°F of Gulf oil standard which has the same viscosity at 210 °F as that of the oil under test

V_x = Viscosity of the oil under test.

V_H = Viscosity at 100°F Pennsylvanian standard oil which has the same viscosity at 210°F as that of the oil under test.

Thus, the higher the viscosity index the lower the rate at which its viscosity decreases with increase of temperature. Hence, oils of high viscosity index i.e., those having flat-viscosity temperature curves are demanded for air-cooled internal combustion engines and aircraft engines. In general, oils of high specific gravity have steeper viscosity-temperature curves. However all oils tend to attain same viscosity above 300°C.

By and large, light oils of low viscosity are used in plain bearings for high-speed equipment such as turbines, spindles and centrifuges whereas high viscosity oil are used with plain bearings of low speed equipment.

The Redwood Viscometer

The Redwood Viscometer is made in two sizes. The Redwood 1 viscometers is commonly used for determination of viscosities of lubricating oils and has an efflux time of 2,000 seconds or less. The Redwood 2 viscometer is similar to the 1 type but the jet for the outflow of the oil is of a larger diameter and hence gives an efflux time of approximately 1/10th of that obtained with 1 instrument under otherwise identical experimental conditions. Redwood 2 instrument is, therefore, used for the oils having higher viscosities, such as the fuel oils.

The Redwood viscometer does not give a direct measure of viscosity in absolute units but it enables the viscosities of oils to be compared by measuring the time of efflux of 50 ml of oil through the standard orifice of the instrument under standard conditions. The results given by these two viscometres are reported as "Redwood 1 Viscosity" or "Redwood 2 Viscosity" followed by the efflux time in seconds at the experimental temperature.

DESCRIPTION

The Redwood I viscometer shown in Fig. 1 essentially consists of a standard cylindrical oil cup made of brass and silvered from inside and has 90 mm height and 46.5 mm in diameter. The cup is open at the upper end. It is fitted with agate jet in the base. The diameter of the orifice is 1.62 mm and the internal length is 10 mm. The upper surface of the agate is ground concave depression into which a small silver plated brass ball attached to a stout wire can be placed in such a way that the channel is totally closed and no leakage of the oil from the cup through the irifice can take place. The cup is provided with a pointer

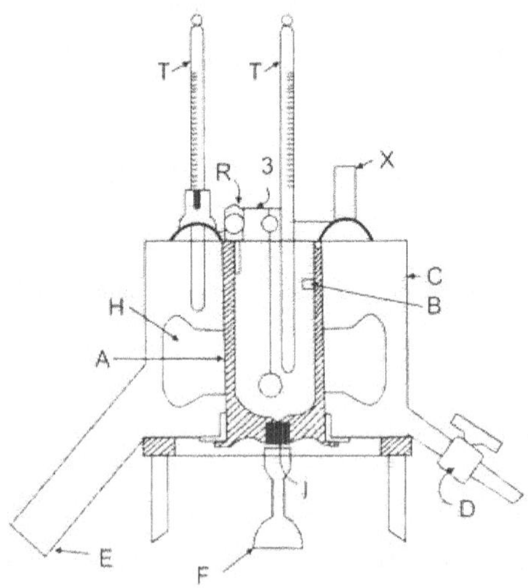

A – Oil cup. B – Gauge Wire. C – Heating bath. D – Water outlet E – Heating tub. J – Agate jet. K – Stirrer. R – Holder. S – Clip. T – T_1 – Thermometers. V – wire. H – Stirrer. F – Flask

Fig. 1. The Redwood No. 1 viscometer

which indicates the level upto which the oil should be filled in the cup. The lid of the cup is provided with The oil cup is surrounded by a cylindrical copper vessel containing water which serves as a water bath to maintaining the desired oil temperature with the help of electrical heating coils of by means of a gas burner as the case may be. A thermometer is provided to measure the temperature of water. A stirrer with four blades is provided in the water bath to maintain uniform temperature in the bath and, hence enabling uniform heating of the oil. The stirrer contains a broad curved flange at the top to act as a shield for preventing any water splashing into the oil cylinder. The entire apparatus rests on a sort of tripod stand provided with levelling screws at the bottom of the three legs. The water bath is provided with an outlet for removing water as and when needed. A spirit level used for levelling the apparatus and a 50 ml flask for receiving the oil from the jet outlet are also provided.

AIM

To determine the viscosity of lubricating oil at different temperatures by redwood viscometer

APPARATUS

1. Redwood viscometer
2. Water bath
3. Stop watch
4. Thermometer

CHEMICALS

1. Lubricating oil

PROCEDURE

(i) Level the instrument with help of the levelling screws on the tripod.

(ii) Fill the water bath with water to the height corresponding to the tip of the indicator up to which the oil is filled in the cylindrical cup.

(iii) Keep the brass ball in position so as to seal orifice.

(iv) Pour the oil under test carefully into the oil cup upto the tip of the indicator.

(v) Keep the 50 ml flask in position below the jet. Keep the oil and water well stirred and note their temperatures.

(vi) When the temperature of the oil and water are steady, raise the ball valve and suspend it from the thermometer bracket. Simultaneously start a stopwatch.

(vii) When the level of oil dropping in to the flask reaches the 50 ml mark, stop the watch and note the time in seconds.

(viii) Replace the ball valve in position to seal the cup to prevent overflow of the oil.

(ix) Refill the oil up to the indicator tip of the oil cup. Repeat the experiment to get nearly reproducible results. Repot the mean value as "Redwood I Viscosity at T°C = t second". This is the viscosity at room temperature, T° C.

(x) Repeat the experiment at five elevated temperatures, say 45°C, 55°C, 65°C, 75°C and 85°C, and note the respective times of efflux as described above. It is convenient to determine the viscosity at the higher temperature first, and then coming down to the lower temperatures.

(xi) For instance, for determining the viscosity at 85°C, raise the temperature of the water bath to about 90°C with the help of the heating coil. Switch off the heating coil. After some time, the oil temperature reaches 85°C.

(xii) Then bring the water temperature to about 86°C by running out some out some water through the out let and adding cold water top while stirring constantly.

(xiii) When the oil temperature is steady, note the time of efflux. Now reduce the temperature water bath by adding more cold water and bring the oil temperature down to 75°C. Repeat the experiment as described above. Similarly determine the efflux times for other lower temperatures.

(xiv) Construct graphs correlating viscosity Vs temperature

CALCULATIONS

Tabulate the result as follows and plot the graphs viscosity Vs temperature

S.No.	Temperature °C	Viscosity, t (seconds, Redwood No. 1)			
		Trial 1	Trial 2	Trail 3	Average
1	Room temp.				
2	75				
3	65				
4	55				
5	45				
6	35				

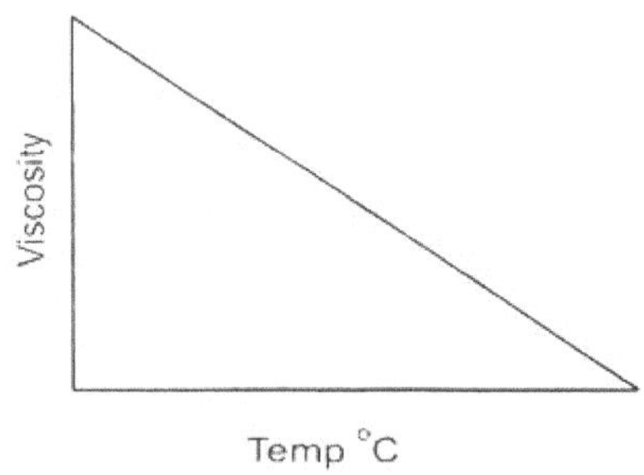

RESULT

From the graph we can observed that as a temperature increases. The viscosity of the lubricating oil decreases.

Example: An oil sample, under test has a Seybolt universal viscosity of 64 seconds at 210°F and 774 seconds at 100°F. The low viscosity standard (Gulf oil) possesses a Seybolt universal viscosity of 64 seconds, at 210°F and 77 seconds at 100°F. The high viscosity standard (Pennsylvantion oil) gave the Seybolt Universal Viscosity Values of 64 seconds at 210°F and 414 seconds at 100°F. Calculate the viscosity index of the oil sample under test.

Solution:

$$\text{Viscosity index of the oil under test} = \frac{V_L - V_X}{V_L - V_H} \times 100$$

$$= \frac{(774 - 564)}{(774 - 414)} = \frac{210}{360} \times 100$$

$$= 58.33$$

EXPERIMENT – 38 Determination of Molecular Weight of a Polystyrene Sample by Viscometric Method

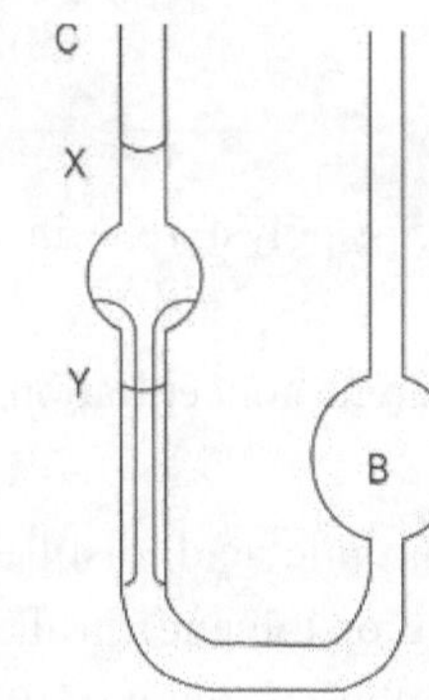

Fig. 1 Ostwald's viscometer (U-tube viscometer)

AIM

To determine the molecular weight of a polystyrene sample by Ostwald's Viscometer.

APPARATUS

1. Viscometer 2. Stopwatch 3. Thermostat at 25 °C

CHEMICALS

1. Polystyrene 2. Benzene 3. Toluene

THEORY

The viscosity of a polymer solution is related to the molecular weight, M. of the polymer by the equation:

$$\frac{\eta_{sp}}{C} = \frac{\eta}{\eta_o} - 1 = KM^{\alpha} \qquad .....(1)$$

where η_{sp} is the specific viscosity of the solution, η_a, and η the viscosities of solvent and solution respectively. C the weight of the polymer (g) in 100 ml of solution, and k and α are constants. This equation is only valid for low concentrations (< 0.50%) of a linear polymer, dissolved in a good solvent, where there is no association. It is thus necessary to extrapolate the curve of C against η_{sp}/C to C = 0; the intercept is the intrinsic viscosity $[\eta]$. Thus, in the limit, eq. 1 becomes

$$[\eta] = KM^{\alpha} \qquad .....(2)$$

In general, the value of K depends on the type of polymer the solvent, and the temperature, while α is a function of the geometry of the molecule. The viscometric method of determining molecular weight is very convenient and gives accurate results if K and α are known. These are obtained from samples of polymers of known molecular weight, in which the molecular weight has been determined by alternative methods.

PROCEDURE

(i) Prepare a solution containing about 0.5 g polystyrene in 100 ml of the solvent. This is 0.5% solution of polystyrene.

(ii) Similarly prepare the solutions of different concentrations (0.1%, 0.2%, 0.3%, and 0.4%) by diluting the above solution.

(iii) Wash the Ostwald's viscometer with chromic acid, distilled water, and dry in an oven. Pipette out 20 ml of the pure solvent (benzene or toluene) in the bulb B of the viscometer which is clamped vertically in a stand, placed in thermostat at 25°C or 30°C (Fig. 1(a)).

(iv) Through a rubber tube attached to upper arm of bulb A, suck up the solvent until it rises above the mark X.

(v) Allow the solvent to fall freely through a capillary up to the mark X. Start the stop watch and note the time t_1 for the flow of liquid from mark X to mark Y.

(vi) Repeat to get five readings and take the mean as the flow time t_0. Remove the solvent, clean the viscometer, and dry it.

(vii) Pipette out 20 ml of one of the solutions prepared above (say 0.5%) and determine the flow time as above. Repeat to get three readings and take their mean.

(vii) Similarly, determine the flow times for solutions of different concentrations after proper cleaning and drying of the viscometer after each set of readings.

OBSERVATIONS

Temperature of the experiment = ———

Solvent used = ———

Value of constant 'K' for styrene/solvent at the above temperature (from the table) = ———

Value of constant 'α' (from the table) = ———

S.No.	Concentration of Polymer Solution	Density	Time of flow	$\frac{\eta}{\eta_o} = \frac{t}{t_o}$	$\eta_{sp} = \frac{\eta}{\eta_o} - 1$	$\eta_{red} = \eta_{sp}/C$
1.	Pure solvent					
2.	0.1%					
3.	0.2%					
4.	0.3%					
5.	0.4%					
6.	0.5%					

CALCULATIONS

1. Plot a graph between η_{red} and concentration and extrapolate the graph to zero concentration (Fig. 2).

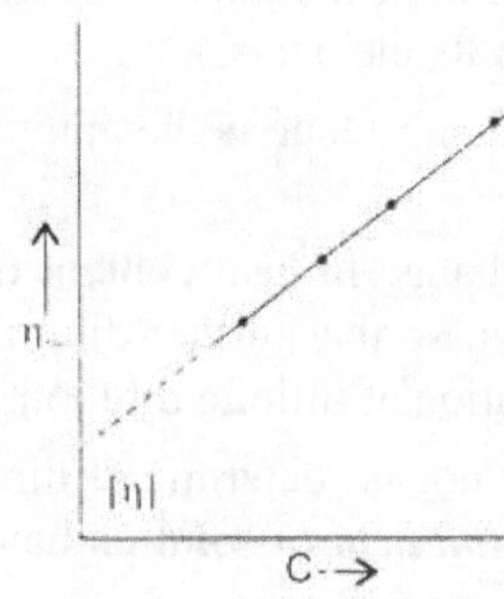

Fig. 2

2. Find out the value of intrinsic viscosity [η] from the graph.

$$[\eta] = \text{Intercept}$$

3. Now we know, $[\eta] = KM^{\alpha}$, substitute the values of [η], K, and α and calculate M, the molecular weight of the polymer.

Using logarithm,

$$\log [\eta] = \log K + \alpha \log M$$

$$\log M = \frac{\log [\eta] - \log k}{\alpha}$$

RESULT

The molecular weight of the given polymer sample is found to be ———

EXPERIMENT – 39 Determination of Heat of Reaction

INTRODUCTION

Chemical reactions are generally accompanied by some heat effects – evolution or absorption of heat. Those taking place with the evolution of heat, are known as *exothermic* whereas those occuring with the absorption of heat, as *endothermic* changes.

The magnitude of heat effect produced in a chemical change, or the *heat of reaction* depends upon (a) pressure, (b) temperature, (c) physical state of the reactants and the products, (d) the amounts of the substances involved. For a particular chemical change, it has a constant value provided the conditions (a) to (d), under which the reaction takes place, remain unchanged. In order to indicate the nature of the reaction different names are given to the heat of reaction, *e.g.* heat of formation, heat of combustion, heat of neutralization, etc.

Heat of formation: It is defined as the change in heat content, *i.e.* amount of heat evolved or absorbed when 1 mole of the substance is formed from its elements.

Heat of combustion: Heat of combustion of a substance is the amount of the heat evolved when 1 gm molecule of it is completely oxidized.

Heat of solution: It may be defined as the change in heat content of the system when 1 mole of the substance is dissolved in large excess of water, so that further dilution of the solution prodces no heat change. It is sometimes known as heat of solution at infinite dilution.

As the heat of solution stated above can not be determined directly, two more heats of solution – (i) *integral heat* of solution and (ii) *differential heat* of solution have been introduced.

Integral heat of solution at a particular concentration *is the heat change when 1 mole of solute is dissolved in pure solvent to produce a solution of the given concentration.* Thus if 1 mole or solute is dissolved in 500 g of water, the heat change gives the value of integral heat of solution at the concentration of 2 molal.

As is obvious from the definitions the integral and differential heats of solution approach each other with decreasing the concentration and become identical when the concentration is zero.

Heat of Hydration: It is defined as the heat change accompanying the formation of hydrats from one mole of anhydrous slat. By measuring integral heat of solution of a salt in its hydrates as well as anhydrous state, it is possible to calculate the heat of hydration of the salt. If ΔH_1 and ΔH_2 are the integral heats of solution of anhydrous and the hydrated salt respectively, then heat of hydration

$$\Delta H = \Delta H_1 - \Delta H_2$$

In this experiment, the heat transfer of reaction [1] will be measured using a solution calorimeter. As the calorimeter operates at *constant pressure*, the heat transfer q_p is equivalent to ΔH for the change of state in [1]. In this way, ΔH and the enthalpy of formation $\Delta H_f^{\circ}(MgO(s))$ will be determined. The limiting reagent in this experiment is 0.5 M HCl.

The solution calorimeter (Fig. 1) consists of a rotary glass sample cell (iii), which doubles as a stirrer. The Teflon sample dish (vii) seals the bottom of this glass sample cell, which can deliver liquid samples up to 20 ml and solid samples up to 2 g into the glass dewar (iv). With the press of the glass push rod (i).

the Teflon sample dish detaches from the glass sample cell and the reaction initiates with the contents of the glass dewar. The thermistor probe (ii) records all temperature changes and output voltages to be recorded for later analysis.

AIM

Determine the enthalpy change ΔH for the reaction

$$MgO(s) + 2H^{+} \longrightarrow Mg^{2+}_{(aq)} + H_2O_{(l)} \qquad .....(1)$$

and calculate $\Delta H_f^o(MgO_{(s)})$.

APPARATUS

1. Calorimeter 2. Volumetric flask

CHEMICALS

1. 0.1 M and 0.5 M HCl 2. MgO

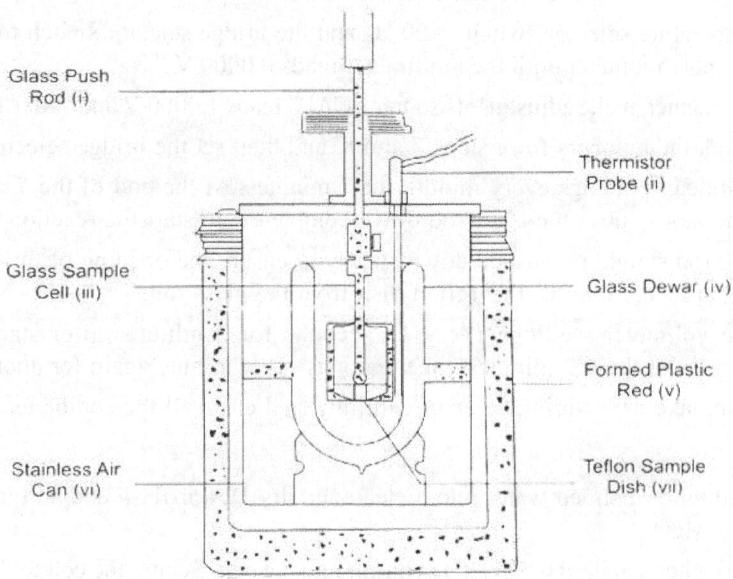

Fig. 1 Parr 1451 Solution calorimeter

PROCEDURE

The heat capacity of the calorimeter must be determined first by supplying it with a *known* quantity of energy derived from a specific reaction. The reaction used for this calibration step will be the protonation of tris (hydroxymethyl) aminomethane [TRIS] with 0.1M HCl.

$$(HOCH_2)_3CNH_2 + HCl \longrightarrow (HOCH_2)_3CNH_3^{+} + Cl^{-} \qquad .....(2)$$

Calibration will be followed by a sample run (*i.e.* reaction [1]). Two sets of data should be acquired, each consisting of a calibration and a sample run.

CALIBRATON

(i) Switch on the power to all of the electrical instruments to allow a 20 minute warm-up time.

(ii) Weigh out 0.50 ± 0.01 g of TRIS into the Teflon dish on the analytical balance and record the mass with ± 0.0001 g accuracy. Ensure that all of the TRIS is inside the dish and insert the dish to the glass bell. This is the sample cell.

(iii) Using a 100.0 ml volumetric flask, add exactly 100.0 ml of 0.1M HCl solution to the Dewar flask. Record the exact molarity of the HCl solution.

(iv) Place the spacer ring on the Dewar and lower the assembly into the air can.

(v) Attach the sample cell to the stirrer shaft and set the unit into the Dewar. Place the thermistor probe into its hole being careful not to jar the sample.

NOTE: The thermistor is very fragile, and quite expensive, so handle with care.

(vi) Start the stirrer motor and set the digital multimeter to DCV and select the 20 V range.

NOTE: On some of the calorimeters, it may be required to place a small weight on the lid of the calorimeter as the stirring motor can cause the lid to jump.

(vii) Set the temperature selector switch to 20 °C and the bridge selector switch to "ZERO". Adjust the variable potentiometer until the multimeter reads 0.0000 V.

(viii) In a similar manner, make adjustments so that "CAL" reads 1.0000 V and "NULL" reads 0.0000 V.

(ix) Double check the numbers from steps 7 and 8, and then set the bridge selector to "READ".

(x) Record voltmeter readings every minute for 7 minutes. At the end of the 7th without delay or stopping the motor, push the glass rod down completely to start the reaction.

NOTE: The rod should be pushed down sternly to ensure the opening of the glass sample cell, but not so hard as to remove the Teflon dish from the glass rod.

(xi) Record the voltmeter readings every 15 seconds for 2 minutes after starting the reaction (8 readings). After those 2 minutes, take readings every minute again for another 8 minutes.

(xii) Stop the run, take the calorimeter apart carefully, and clean all the components.

Samples

(xiii) Pipette 80.0 ml of distilled water into a clean and dry Dewar flask and add it to approximately 0.3-0.4 g of $MgO_{(s)}$.

(xiv) Pipette 20.0 ml of standard 0.5M HCl solution into the cell. Secure the cell to the stirrer shaft and reassemble the apparatus as in the calibration.

(xv) Take readings on the same time scale as with the calibration in steps 10-12.

(xvi) Stop the run, take the calorimeter apart carefully, and clean all the components.

(xvii) Repeat steps 1 through 16.

CALCULATIONS

Convert all of the voltmeter readings to degrees °C. When the thermistor bridge is adjusted in the manner described in the procedure, the voltmeter reading is related to the temperature in °C as follows:

$$(\text{Voltage reading}) \times 10 = \text{temperature in } ^\circ C \quad .....(3)$$

If the selector switch is set to 20 °C, then a voltmeter reading of 0.3459 V would correspond to 20 °C + 3.459 °C = 23.459 °C.

Construct four thermograms with the 2 calibration and 2 sample runs. Plot temperature versus time and determine from them the (corrected for heat losses/gains) temperature changes ΔT_c for all the reactions. Determine $T_{0.63R}$ from the plots as well.

Calculate the heat capacity by calculating the energy input from the reaction:

$$Q_E = m\,[245.760 + 1.4364\,(25 - T_{0.63R})] \quad .....(4)$$

where Q_E is the energy input in Joules

m is the mass of TRIS in grams

$T_{0.63R}$ is the temperature at the 0.63R point on the thermogram

The 1.4364 (25-$T_{0.63R}$) term in (eq. 6) adjusts the heat of reaction to any temperature above or below the reference temperature of 25 °C.

Calculate ΔH of reaction (eq. 3) and consequently the molar enthalpy change based on the number of moles of HCl used. Calculate the enthalpy of formation $\Delta H_f°(MgO_{(s)})$ from your experimental results and from the theoretical enthalpies of formation of $H_2O_{(l)}$ and $Mg^{2+}{}_{(aq)}$ (available in any CRC reference). Estimate quantitatively the limits of uncertainty of your results.

DISCUSSION

Compare the calculated $\Delta H_f°(MgO_{(s)})$ value with the theoretical one and discuss any discrepancies that you may find between them. Discuss in detail any shortcomings of the experimental procedure and set up. Also point out the main sources of error, and offer solutions to minimise their effect on the final results.

Experiment - 40 Determination of Heat of Neutralization of Hydrochloric Acid

INTRODUCTION

Heat of neutralization : The heat evolved when 1-gram equivalent of an acid/base is completely neutralised by a base/acid. When the acid, base and the salt are strong electrolytes and solutions are quite dolute the process of neutralisation can be represented as

$$HA + BOH \rightarrow BA + H_2O + 57.3 \text{ kJ mol}^{-1}$$

or
$$H^+ + A^- + B^+ + OH^- \rightarrow A^- + B^+ + H_2O + 57.3 \text{ kJ mol}^{-1}$$

Cancelling the common ions present on both the sides, we have

$$H^+ + OH^- \rightarrow H_2O + 57.3 \text{ kJ mol}^{-1}$$

It is this seen that the net reaction, which takes place in the process of neutralisation is the combination of hydrogen and hydroxyl ions in the form of water. It follows, therefore that the heat neutralisation of a strong acid againt a strong base is merely the heat of formation of water from hydrogen and hydroxyl ion, which is equal to 57.3 kJ mol^{-1}. Thus the heat of neutralisation of all strong acids versus strong base should base the same.

Look at reaction in the ions A^- and B^+ were present before and after the reaction. H^+ and OH^- ions are present before and after the reaction, from where the heat comes. Let the absolute enthalpy of an acid be H_{HA} and that of base be H_{BOH} while for HBA for salt H_{H_2O} are absolute enthalpy of salt BA and H_2O so that

Heat of neutralisation $= \Delta H + H_{BA} + H_{H_2O} - H_{HA} - H_{BOH}$

If either the acid or base isweak of both are weak, the heat of neutralisation is not 57.3 kJ mol^{-1} but less than this value. This is because the acid or base or both are not in an ionised state. In order to remove the ions from molecular state, some heat is used up in ionising those molecules hence heat of neutralisation less than 57.3 kJ mol^{-1}.

The classical calorimeter: The classical calorimeter due to Ostwald is shown in Fig. 1. The central vessel, *A*, is a metal can, usually of silver, having capacity of 500-800 ml. The inner and outer surface of this vessel are polished and it is surrounded by two or more polished metal radiation screens, *B*, insulated from each other by wooden blocks, *C*. Finally, the apparatus is surrounded by a double walled

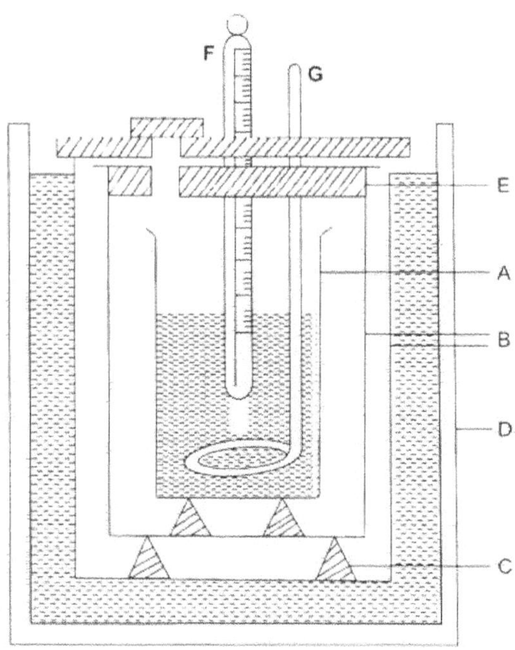

Fig. 1 Calorimeter

jacket, *D*, itself enclosed in felt and filled with water. To prevent the air droughts and evaporation of the liquid in the calorimeter all the compartments are covered with non-conducting ebonite lids, *E*. A sensitive thermometer, *F* and a glass stirrer, *G*, are fitted into the calorimeter through the holes provided in the lids.

AIM

To determine the heat of neutralization of hydrochloric acid using a classical calorimeter.

APPRATUS

1. Classical claorimeter 2. Two 0.1 × 50° thermometer 3. A ring stirrer of glass tubing.

CHEMICALS

1. 0.25N HCl 2. 0.25N NaOH

THEORY

A known volume of HCl solution of exactly known concentration is allowed to react completely with a strong alkali in dilute solution and the rise in temperature is determined. Knowing the heat capacity of calorimeter together with the bulb of the thermometer immersed, the masses of the acid and alkali solutions and their specific heats, the amount of heat liberated can be obtained. With the known volume and concentration of HCl solution, the heat of neutralization can then be calculated. If Q is the amount of heat evolved when V ml of HCl solution of concentration C g equivalents/litre is neutralized completely, then

$$\text{Heat of neutralization} = 1000 \times Q/V \times C$$

PROCEDURE

(i) From the standard HCl prepare 1N HCl solution. Dissolve 40 g NaOH and make up the volume to 1 litre to get ≈ 1N NaOH but it should be more than 1N. Leave both the solutions overnight so that both attain same room temperature.

(ii) It is assumed that the heat capacity and density of 1N HCl is same as that of water for all temperatures. However, normality of acid should be known exactly. Measure 50 ml of ≈ 1N NaOH and place it in thermocole flask with thermometer and stirrer.

(iii) Note the thermometer reading after every half-a-minute interval for five minutes.

(iv) Pour it into the calorimeter beaker containing alkali quickly with constant stirring. Note the exact time of addition and the thermometer readings after every half-a-minute interval for five minutes. In the cource of taking readings if a few readings are missed out put a dash against that time column but do not stop the stopwatch.

(v) Plot a graph of temperature versus time (Fig. 2) and find out the temperature of the mixture at the moment of mixing. For this draw a verticle line for the moment of mixing and extend the temperature-time graph of the pure acid, pure base and mixture to intersect with the vertical line.

(vi) Calculate the heat of reacton. Repeat two more readings for finding out the heat of neutralization.

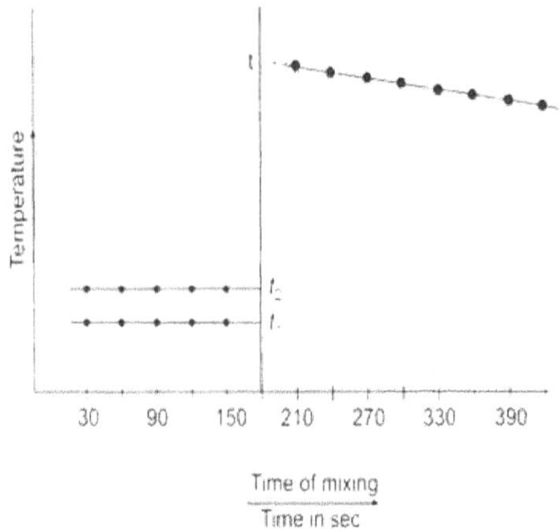

Fig. 2 Temperature versus time plot for HCl versus NaOH reaction

Calculation

Temperature of ≈ 1N NaOH = t_1°C

Temperature of 1N HCl = t_2 °C

Density of water (≈ to 1N NaOH) at $\frac{t_1 + t_2}{2} = \rho$

Heat capacity of water (≈ to 1N NaOH) at $\frac{t_1 + t_2}{2} = C_p$

Total mass of solution = 100 ρ

Temperature of mixture at mixing time = t °C

Heat capacity of calorimeter beaker for 50 hot + 50 cold case = C.

Heat evolved in the reaction calculated as

$$\Delta H = (C + 100\ \rho)\ C_p \left(t - \frac{t_1 + t_2}{2}\right)$$

As heat is evolved in this reaction and 50 m*l* of NHC*l* ≡ N equivalent.

Heat of neutralization, $DH_{neutralization} = -N\ \Delta H$. KJ mol^{-1}

RESULT

Heat of neutralization = ——— KJ mol^{-1}

EXPERIMENT – 41 Adsorption of a Solute on Charcoal

INTRODUCTION

It has been observed that there are a number of porous solids like coconut charcoal, silica gel, fine powders as fuller' s earth and colloids like ferric hydroxide or aluminium hydroxide which have the property of accumulating other substances from surrounding gases and solutions at their surfaces. This phenomenon is known as *adsorption.* As it is a surface phenomenon the amount adsorbed will greatly depend on the specific area of the adsorbing substance. The substance which adsorbs is known as an *adsorbent* and the substance which gets adsorbed is known as an *adsorbate.*

The surface area available per gram of an adsorbent is known as its specific area. This will obviously depend on the porosity and the state of subdivision of the adsorbent.

The amount of gas adsorbed by a given amount of an adsorbent depends both on the prevailing temperature and the pressure. At a given temperature, the variation in amount of gas adsorbed with change in pressure is given by an empirical relation suggested by Freundlich, and known as the *Freundlich adsorption isotherm.* This is:

$$\frac{x}{m} = K_1 P^{\frac{1}{n}} \quad \text{or} \quad \left(\frac{x}{m}\right)^n = K_2 P$$

where x is the weight of gas adsorbed by m grams of adsorbent at an equilibrium pressure P, the two K's and n are characteristic constants whose values depend on the temperature and the nature of the adsorbent and of the adsorbate. For solutions, the above equation is written as:

$$\frac{x}{m} = KC^{\frac{1}{n}}$$

where C stands for the equilibrium concentration of the solute in solution.

Taking logarithms of the terms in the Freundlich isotherm we get,

$$\log \frac{x}{m} = \log K + \frac{1}{n} \log C$$

If we plot a graph between log (x/m) (along ordinate) versus log C (along abscissa), a straight line would result. The slope of this line will equal $1/n$ and log K can be obtained as the intercept which the straight line makes on the ordinate for a value of log C equal to zero.

AIM

To verify the Freundlich isotherm for adsorption of acetic acid on a given sample of charcoal.

APPARATUS

1. Pipette
2. Burette
3. Six 100 ml conical flasks
4. Six funnels
5. Six pieces of glazed paper 5 cm × 5 cm.

CHEMICALS

1. M/2 acetic acid   2. M/10 NaOH   3. Charcoal

PROCEDURE

(i) Standardise the solutions of acetic acid and sodium hydroxide.

(ii) Take six dried and cleaned 100 ml conical flasks. Number them from 1 to 6. Set up two burettes, one for water and the other for M/2 acetic acid solution.

Flask number 1 → 50 ml acetic acid + 0 ml water

Flask number 2 → 40 ml acetic acid + 10 ml water

Flask number 3 → 30 ml acetic acid + 20 ml water

Flask number 4 → 20 ml acetic acid + 30 ml water

Flask number 5 → 10 ml acetic acid + 40 ml water

Flask number 6 → 0 ml acetic acid + 50 ml water

(iii) Mix the solutions Well.

(iv) To each of the flasks, add two grams powdered charcoal in quick succession. (For this six different two gram quantities of adsorbent charcoal can be kept weighed on glazed papers under watch glasses or funnels for protection against air currents).

(v) Shake well and place the flasks in a thermostat at room temperature for about 30 minutes. Shake the contents from time to time.

(vi) Calculate the initial concentrations of acid for the five flasks and correct them for the titrant reading for the Sixth (blank) flask.

(vii) Set up six funnels fitted with dry filter papers of equal size and of good quality. Reject about 5 ml of initial filtrate in each case and collect the remaining titrates in separate conical flasks.

(viii) Titrate 10 ml of each solution against standard alkali. Tabulate the data as suggested further on.

Temperature = ...°C

Mass of charcoal, m = 2 g (per bottle)

10 ml of the original M/2 acid required 50 ml of M/10 NaOH

Flask nos.	ml of M/10 NaOH required for 10 ml solution after adsorption	Initial concentration of acetic acid, C_i in moles/litre	Equilibrium concentration of acetic acid in mole/litre	Amount adsorbed $x = \frac{C_i - C_e}{20}$	x/m	$\log \frac{x}{m}$	$\log C_e$
1							
2							
3							
4							
5							

CALCULATIONS

First calculate the concentration (C_i) of acetic acid in moles/litre for each flask. From the titration data, obtain the equilibrium value of C_e for each flask. The value of x for each flask can be calculated from the change in concentration due to adsorption i.e., ($C_i - C_e$) and the initial volume of the solutions (50 ml) as follows:

$$x = \frac{(C_i - C_e)50}{1000} = \frac{(C_i - C_e)}{20} \text{ moles}$$

From this, calculate the amount of acid adsorbed per gm of the adsorbent (x/m) in moles or in grams.

1. Plot a graph between log (x/m) values (ordinate) versus log C_e (abscissa). If the plot obtained is a straight line, this will prove the validity of the Freundlich isotherm over the concentration range considered. Evaluate the value of $1/n$ from the slope of the line and log K from the intercept on the ordinate by the straight line for a value log C_e equal to zero.

RESULT

The plot obtained is a straight line, so the Freundlich isotherm is verified.

Experiment - 42 Photo Chemical Reduction of Ferric Salt (Blue - Printing)

INTRODUCTION

A study of the chemistry involved in the printing and developing of blue-prints is an instructive experiment. In addition to the electrochemical aspects. the experiment provides an opportunity to study a photochemical reaction.

Ferric ions are reduced by oxalic acid to the ferrous state and the rate of reduction is enhanced by exposing the reactants to light. When ferrous ions react with ferricynaide ions. a blue color (Turnbull's blue) is developed. The quantity of ferric salt reduced to the ferrous state under the influence of light. is made apparent by the depth of the blue color.

Since the ferric oxalate oxidation-reduction reaction is very fast in light. diammonium phosphate us added to reduce the sensitivity of the ferric oxalate reaction so that the sensitized paper may be prepared the diffused light of the laboratory. If diammonium phosphate is not added. the use of a dark room is essential in order to prepare sensitive blue-print paper. A very dilute solution of potassium dichromate (0.03M) is used to wash the finished print.

In the printing operation. a negative of an object. whose silhoutte is to be printed. is placed on the sensitive paper and the uncovered portion is exposed to sunlight. Finally. the paper is treated with a $K_3Fe(CN)_6$ solution and the excess of unreacted ferric oxalate is washed away with water.

$$2Fe^{+3} + C_2O_4^{-2} \rightarrow 2Fe^{+2} + 2CO_2$$

$$3Fe^{+2} + 2[Fe(CN)_6]^{-3} \rightarrow Fe_3[Fe(CN)_6]_2$$

AIM

To study the photochemical reduction of a ferric salt (Blue-printing)

APPARATUS

1. Beaker
2. Glass rod

CHEMICALS

1. 0.5M oxalic acid
2. 0.67M ferric chloride
3. 0.1M potassium ferricyanide
4. 3.5M diammonium phosphate
5. 0.03M potassium dichromate
6. 0.1M hydrochloric acid

PROCEDURE

Prepare the above solutions in water.

(i) In a 400 ml beaker. mix 100 ml of oxalic acid solution with 20 ml of diammonium phosphate solution. Place the beaker in diffuse light (in your locker).

(ii) Add 100 ml of ferric chloride solution to the solution of oxalic acid and diammonium phosphate under stirring (in diffuse light). A small precipitate formed initially will dissolve on further stirring. Close your locker and open it only when needed. This solution contains ferric oxalate.

(iii) Take four pieces of typewriter paper (4" × 2.5"). Open your locker and immerse the papers in the sensitizing solution of ferric oxalate. This is best done by curling the pieces of paper in a semi-cylindrical way and sliding them into the beaker. Stir the solution or rotate the beaker so that the papers are thoroughly wet and no dry spot are left. (This should be done in diffuse light inside your locker).

(iv) Remove the wet pieces of paper from the beaker and place them between sheets of filter paper. This should be done as quickly as possible in the partially closed locker. Leave the pieces of paper between the filter paper sheets for 15-20 minutes so that they may get sufficiently dry. If a sharp print is needed, the papers should be dried over-night in the locker, otherwise the edges of the print wil be fuzzy.

(v) After the papers have dried sufficiently, take one of them and place the opaque object, or a negative on the top of the sensitized paper, compress it between two sheets of glass and expose it to light. The time of exposure for a normal printing is four to five minutes, if bright sunlight is used. While printing, do not hold the glass plates but lay them on the desk, or any other flat surface.

(vi) Dip the paper after exposure to light, into a 0.1M $K_3Fe(CN)_6$ solution kept in a 400 ml beaker. This and the following operations may be carried out in diffuse light on the top of the desk. It is important that the paper be immersed all at once, otherwise lines will occur in the blue field of the print. Remove the paper and dip it in the solution of potassium dichromate. Wash the paper, first in 100 ml of 0.1M hydrochloric acid, then in tap water.

(vii) Use the other pieces of sensitized paper to make a series of exposures of the same object varying the time of exposure in order to obtain the most satisfactory print.

(viii) Mount the blue prints, which you have made, in your note book and indicate the time of exposure in minutes for each print.

Part - III Organic Chemistry Experiments

Identification and Preparation of Organic Compounds

EXPERIMENT – 43 Identification of Functional Groups Present in Known/Unknown Organic Compound

PRELIMINARY EXAMINATION

1. State 2. Colour 3. Odour/Smell 4. M.P/B.P

SOLUBILITY TEST

1. Water 2. Ether 3. Sodium Hydroxide solution
4. Sodium Bicarbonate solution 5. Dilute HCl 6. Conc.H_2SO_4

IGNITION TEST FOR AROMATIC OR ALIPHATIC NATURE

Burn small amount of the substance on a given spatula. If the compound burns with a sooty flame it is aromatic. Otherwise it is aliphatic compound.

UNSATURATION TEST

1. *Bromine Test:* Dissolve small amount of the substance in water or alcohol. To this add 1-2 drops of bromine water and shake. Discharge of colour indicates the presence of unsaturation
2. *Bayer's Test:* Dissolve a small amount of the substance in about 5 ml of water or alcohol. To this add 2-3 drops of Bayer's reagent. If the purple colour disappears indicates the presence of unsaturation.

There is a possibility for the presence of Extra elements other than functional groups. So here by these tests we can find out the extra elements present in the given substance.

Extra elements test/Sodium fusion extract test/Lassaigne's test

A Small piece of fresh Sodium metal is taken in a fusion/ignition tube and it is heated under Bunsen burner until it becomes red hot, then add small amount of the substance is added in that ignition/fusion tube and again heated until red hot. Add this ignition tube in an mortar pestle containing 5 ml of distilled water, grind the tube and filter it. Divide the fusion extract into 3 parts and test it for Nitrogen, Halogens and Sulphur.

Test for Nitrogen

To one part of the sodium fusion extract add 1ml of freshly prepared ferrous sulphate and boil for 2 minutes and add 2 drops of conc. sulphuric acid along the walls of the test tube.

If the solution turns Prussian blue. It indicates the presence of Nitrogen.

$$FeSO_4 + 2\,NaOH \longrightarrow Fe(OH)_2 + Na_2SO_4$$

$$Fe(OH)_2 + 6NaCN \longrightarrow \underset{\text{Sodium Ferrocyanide}}{Na_4[Fe(CN)_6]} + 2NaSO_4$$

$$3Na_4[Fe(CN)_6] + 2[Fe_2(SO)_4] \longrightarrow \underset{\text{Ferri-Ferrocyanide}}{Fe_4[Fe(CN)_6]_3} + 6Na_2SO_4$$

Test for Halogens

To the second part of the sodium fusion extract add dil.nitric acid and add freshly prepared silver nitrate solution.

If the solution turns to curdy white precipitate it indicates the presence of Chlorine.

If the solution has light yellow colored precipitate it indicates the presence of Bromine.

If the solution has Dark yellow colored precipitate it indicates the presence of Iodine.

$$Na^{+}X^{-} + AgNO_3 \xrightarrow{\text{dil } HNO_3} AgX + NaNO_3$$

Test for Sulphur

To the third part of the sodium fusion extract add dilute acetic acid and few drops of lead acetate solution.

A black/brown precipitate of lead sulphide indicates the presence of sulphur.

$$Na_2S + 2CH_3COOH \longrightarrow 2CH_3COO^{-}Na^{+} + H_2S$$

$$(CH_3COO)_2Pb + H_2S \longrightarrow PbS + 2CH_3COOH$$

Test for identification of fuctional groups

1. Test for carboxylic acids

Solubility in

Water	Ether	$NaHCO_3$	NaOH	dil. HCl
(-)	(+)	(+)	(+)	(-)

(a) Dissolve the substance in aqueous $NaHCO_3$ (provided), if bubbles of carbon dioxide are seen it indicates presence of acid group (some phenols also give this test).

(b) If the substance is soluble in water, prepare an aqueous solution and check its pH (use pH paper). If the substance is insoluble in water, dissolve in alcohol and check the pH.

(c) Warm a small amount of the substance with two parts of absolute ethanol and one part of con. H_2SO_4 for 5 minutes; cool and pour the mixture slowly in to aqueous Na_2CO_3 solution contained in evaporating dish , and smell immediately. An acid usually yields a sweet, fruity smell of an ester (acids of high molecular weight often give almost odorless esters).

Derivative for Carboxylic acids

Amides: To a small portion of Substance and Thionyl chloride as it converts to acid chloride add ammonia solution. It yields amide.

$$RCOOH + SOCl_2 \longrightarrow RCOCl + SO_2 + HCl$$

$$RCOCl + 2NH_3 \longrightarrow RCONH_2 + NH_4Cl$$

2. Test for Phenols

Solubility in

Water	Ether	$NaHCO_3$	NaOH	dil. HCl	con. H_2SO_4
(-)	(+)	(-)	(+)	(-)	(+)

(a) *Ferric Chloride Test:* Add few drops of aqueous ferric chloride solution to a small amount of the substance. Most phenols produce an intense red, blue, purple or green color due to complex formation.

$$6\,AR\text{-}OH + FeCl_3 \longrightarrow 3H^+[Fe(O\text{-}AR)_6]^{3-} + 3HCl$$

(b) *Liebermann test :* Place 0.2 gm of the substance in a dry test tube and dissolve in 1 ml of conc. sulphuric acid. Add a few crystals of sodium nitrite, immediately a blue green/blue violet colour is formed which on dilution with water it turns red, which on further dilution with sodium hydroxide solution it turns blue. Color formation is observed due to the formation of Iodophenol.

Derivative for Phenols

Benzoates : Place 0.5 ml of the phenol in 2.5 ml of water in a test tube. To this 2.5 ml of 10% sodium hydroxide solution is added followed by the addition of 0.3 ml of benzoyl chloride. Stopper the tube and shake for several minutes. The odor of benzoyl chloride should disappear. Collect the solid on a Buchner funnel.

3. Test of Aldehydes and Ketones

Solubility in

Water	Ether	$NaHCO_3$	NaOH	dil. HCl	con. H_2SO_4
(–)	(+)	(–)	(–)	(–)	(+)

(a) *2, 4-Dinitrophenylhydrazine test:* To small amount of substance dissolved in ethanol, 1ml of 2, 4-dinitrophenlhydrazine reagent is added and shakes the mixture vigorously. Yellow/orange colored precipitate indicates the presence of aldehydes and ketones.

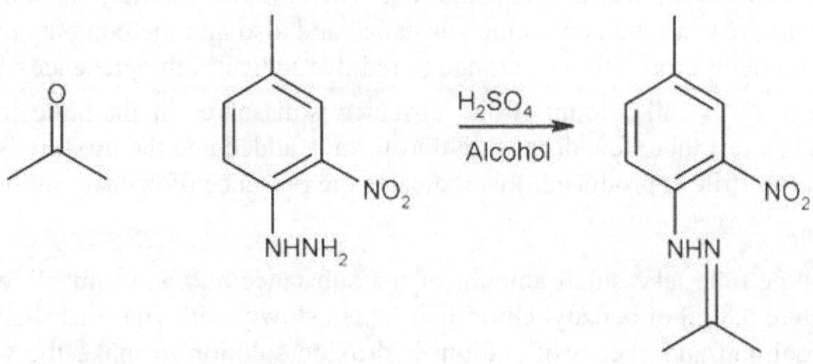

(b) *Fehling's test:* To a small amount of the substance Fehling's Solution A and Fehling's Solution B are added in equal amounts. A reddish brown precipitate indicates the presence of aldehydes

(c) *Tollen's test:* To small amount of the substance add 2 or 3 ml of freshly prepared Tollen's reagent and heat the mixture in water bath. If a silver mirror is deposited on the inner walls of the test tube, it indicates the presence of aldehydes. Ordinary ketones do not give positive result in this test.

Derivative for aldehydes and ketones

Semicarbazones: Place 0.5 gm of semicarbazide hydrochloride, 0.8 gm of sodium acetate, 5 ml of water and 1ml of ethanol in a test tube. To this add a small amount of the substance. Shake the mixture for few minutes then warm the solution in the water bath for 15 minutes. And allow to cool. The semicarbazone precipitates out from the cold solution on standing.

Test for Carbohydrates

Solubility in

Water	Ether	$NaHCO_3$	NaOH	dil. HCl	con. H_2SO_4
(+)	(–)	(+)	(+)	(+)	(+)

(a) *Molisch Test:* To a small amount of the substance 10% solution of α-naphthol in alcohol is added and add few drops of conc. sulphuric acid drop wise without disturbing the test tube. Formation of purple ring at the junction of the layers indicates the presence of carbohydrates

(b) *Benedict's test:* Add 2 or 3 ml of the Benedict's reagent to the small amount of the substance taken in a test tube. Boil the contents for 2-3 min. A red, brown or yellow precipitate indicates the presence of the reducing sugar.

Derivative of carbohydrates

Osazone: A small amount of the substance is taken in a test tube, to this add 0.4 g of phenylhydrazine and 0.6 gm of sodium acetate and 4 ml of distilled water. Heat the test tube in the waterbath for 30 min. Yellow or brown colored precipitate indicates the formation of osazone.

Test for amines

Solubility in

Water	Ether	$NaHCO_3$	NaOH	dil. HCl	con. H_2SO_4
(–)	(–)	(–)	(–)	(+)	(+)

1. *Diazotization and coupling test:* To a small amount of the substance add few ml of 2N hydrochloric acid and cool the solution at 5° in an ice bath. Cool the second test tube containing small amount of α-naphthol in 10% sodium hydroxide solution. When the two test tubes are cooled add sodium nitrite solution to the test tube containing substance and also add the contents of the second test tube. Shake thoroughly: formation of orange to red dye indicates the presence of primary amine.
2. *Carbylamine's test:* A small amount of the substance is dissolved in alcoholic sodium hydroxide solution taken in a test tube, few drops of chloroform is added and the mixture is warmed gently. A foul odor of isonitrile is produced; this indicates the presence of primary amine.

Derivative for amines

Benzamide: In a boiling tube take small amount of the substance and add 3ml of sodium hydroxide solution , to this mixture 0.8 ml of benzoyl chloride is added slowly with vigorous shaking. Heat it on a steam for 30 min, cool and add excess of sodium hydroxide solution to make the solution alkaline. Benzamide is separated as solid.

Test for Hydrocarbons

Solubility in

Water	Ether	$NaHCO_3$	NaOH	dil. HCl	Conc. H_2SO_4
(–)	(–)	(–)	(–)	(–)	(+)

(a) *Formalin test:* Small amount of the substance is dissolved in carbon tetrachloride and this is added to the test tube containing drops of formalin in 2 ml of conc. sulphuric acid.

Dark red color indicates the presence of Benzene (b.p. 110°) or Naphthalene (m.p.80°).

Dark brown color indicates the presence of o-xylene (b.p.144°).

Dark yellow color indicates the presence of Anthracene (m.p.216°).

(b) *Aluminium chloride:* To a small amount of the substance dissolved in chloroform, a pinch of anhydrous aluminium chloride along the sides of the test tube.

If yellow colored $AlCl_3$ crystals are obtained and soon turn to dark orange it indicates the presence of benzene.

If it appears green color in the beginning and turn to blue it indicates the presence of Naphthalene.

If it appears orange color and chloroform layer changes to yellow color it indicates the presence of o-xylene.

If the solution appears light yellow to green color and chloroform layer is colorless it indicates the presence of Anthracene.

Derivative for Hydrocarbons

Picrates: Dissolve equimolar portions of the substance and picric acid in two separate test tubes in 3ml of boiling benzene. Mix the solutions while hot and heat at 50°-60° for 5 min. Allow the mixture to cool, colored precipitate is obtained.

EXPERIMENT - 44 Preparation of Thiokol Rubber

INTRODUCTION

Another important synthetic rubber is thiokol. It is a polymer of ethylene polysulphide. It can be prepared by the condensation of 1,2-dichloroethane with sodium polysulphide.

$$\underset{\text{1, 2dichloroethane}}{Cl-CH_2-CH_2-Cl} + \underset{\text{soidum polysulphide}}{Na-S-S-Na} + \underset{\text{1, 2dichloroethane}}{Cl-CH_2-CH_2-Cl}$$

$$\downarrow \text{Polymerisation}$$

$$\underset{\text{Thiokol [ethylene poly sulphide polymer]}}{Cl-[-CH_2-CH_2-S-S-]_n CH_2-CH_2-Cl} + \underset{\text{sodium chloride}}{2n\,NaCl}$$

PROPERTIES

Thiokol rubber is resistant to the action of oxygen and ozone. It is also resistant to the action of petrol, lubricants and solvents.

USES

- Hoses and tank linings for the handling and storage of oils and solvents.
- Lining of vessels used in the manufacture of chemicals.
- Engine gaskets and other such products that come into contact with oil.
- Thiokol mixed with oxidizing agents is used as a fuel in rocket engines.

AIM

To synthesise Thiokol rubber using sodium poly sulphide with 1, 2 dichloro ethane

APPARATUS

1. Beakers 2. Glass rod

CHEMICALS

1. NaOH
2. Powdered Sulphur
3. Ethylene chloride (1, 2 dichloro ethane)
4. Benzene
5. 5% Sulphuric acid and Nitric acid

THEORY

It is a rubbery white substance and is obtained by treating sodium polysulfide with 1, 2-dichloro ethane.

$$\mathrm{NClH_2CH_2Cl + n\ Na{-}\overset{\overset{S}{\|}}{S}{-}\overset{\overset{S}{\|}}{S}{-}Na}$$

$$\xrightarrow{\Delta} \left[\mathrm{CH_2CH_2{-}\overset{\overset{S}{\|}}{S}{-}\overset{\overset{S}{\|}}{S}}\right]_n + \mathrm{2N\ NaCl}$$

PROCEDURE

(i) In a 100 ml beaker dissolve 2 g NaOH in 50-60 ml of warm water.

(ii) Boil the solution and to this add in small lots with constant stirring 4 g of powdered sulfur. During addition and stirring the yellow solution turns deep-red.

(iii) Cool it to 60-70°C and add 10 ml of 1, 2-dichloroethane (ethylene chloride) with stirring. Stirr for an additional period of 20 minutes while rubber polymer separated out as a lump.

(iv) Pour out the liquid from the beaker in the sink to obtain thiokol rubber. Wash it with water under the tap.

(v) Dry in the fold of filter papers. The yield is about 1.5 g. Determine the solubility of the polymer in benzene, acetone, 5% H_2SO_4 and nitric acid.

Note : If some sulfur remains undissolved filter the solution.

RESULT

Yield obtained = ——— g

Melting point = ——— °C

EXPERIMENT – 45 Preparation of Aspirin

INTRODUCTION

Aspirin is a trade name for acetylsalicylic acid. It has long been known as an excellent analgesic, antipyretic (temperature-lowering) and anti-inflammatory agent, and is still finding new uses, e.g. it lowers the risk of heart attacks and strokes. Although its antecedents can be traced to herbal remedies used as far back as ancient Greece, it was first synthesised in pure form in 1897 by chemists at the German company Bayer, and commercial production began in 1899.

Qualitatively, the purity of an aspirin sample can be determined from its melting point. The melting point of a substance is essentially independent of atmospheric pressure, but it is always lowered by the presence of impurities (a colligative property of pure substances). The degree of lowering of the melting point depends on the nature and the concentration of the impurities.

AIM

To synthesize common pain reliever: aspirin and to determine the purity of the aspirin.

APPARATUS

1. Erlenmeyer flask
2. Measuring Jar
3. Beaker
4. Stirring rod
5. Buchner funnel
6. Melting point tube
7. Watch glass

CHEMICALS

1. Salicylic acid
2. Acetic anhydride
3. Acetic acid
4. Ethanol
5. Ferric chloride solution

THEORY

Aspirin is the acetyl ester of a phenol called salicylic acid. Acetylation of an alcohol or phenol can be carried out with a variety of acid or base catalysts, and in this experiment we will use acetic acid. It is important that the apparatus should be scrupulously dry because a small amount of water will reverse the reaction, to reform the reactant. The product, a typical organic compound, is insoluble in water and is isolated by precipitation from water.

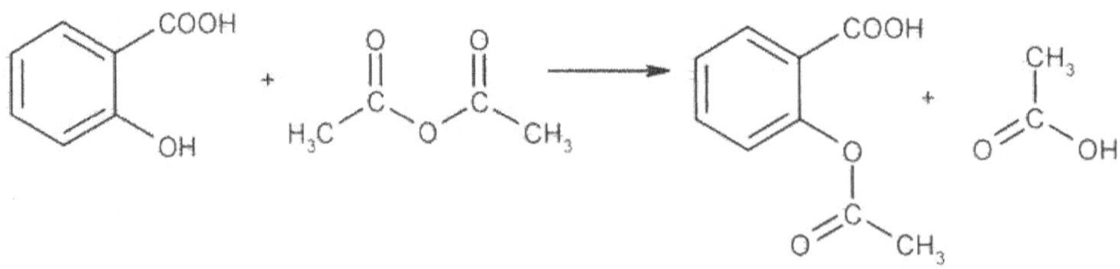

Salicilic acid Acetic anhydride Acetylsalicylic acid

EXPERIMENTAL PROCEDURE

(i) Weigh out 2.0 g of salicylic acid. Place it in a 125 ml Erlenmeyer flask.

(ii) Add 5 ml of acetic anhydride. Swirl the flask to wet the salicylic acid crystals. Add 5 drops of concentrated sulfuric acid, H_2SO_4, to the mixture.

(iii) Gently heat the flask in a boiling water bath for about 10 minutes.

(iv) Remove the flask from the hot water bath and add 10 ml of deionized ice water to decompose any excess acetic anhydride. Chill the solution in an ice bath until crystals of aspirin no longer form, stirring occasionally to decompose residual acetic anhydride. If an "oil" appears instead of a solid, reheat the flask in the hot water bath until the oil disappears and again cool.

(v) Set up a vacuum filtration apparatus. Wet the filter paper in the Buchner funnel with 1-2 ml of distilled water.

(vi) Turn on the water aspirator. Decant the liquid onto the filter paper, minimizing any transfer of the solid aspirin. If some aspirin is inadvertently transferred to the filter, that will not cause any difficulty.

(vii) Add 15 ml of cold water to the flask, swirl, and chill again. Pour the liquid and the crystals of aspirin onto the filter paper. Repeat until the transfer of the crystals to the vacuum filter is complete. Wash the aspirin crystals on the filter paper with 10 ml of ice water.

(viii) Maintain the vacuum to dry the crystals as best possible.

(ix) If aspirin forms in the filtrate in the vacuum flask, transfer the filtrate and aspirin to a beaker, chill in an ice bath, and vacuum filter as before, using a new piece of filter paper. Dispose of the filtrate in the sink.

(x) Determine the mass of the crude aspirin crystals.

RECRYSTALLIZATION OF THE ASPIRIN

The major impurity in aspirin is salicylic acid. It can be removed by a recrystallization.

(i) Place the aspirin crystals in a 100-ml beaker. Add 8 ml of ethanol and 25 ml of water.

(ii) Warm the mixture in a 60°C water bath (no flame, use a hot plate or a hot water bath). Warm the mixture until the aspirin dissolves. (If the solid does not dissolve after heating, consult with your instructor.)

(iii) Cover the beaker with a watch glass, remove it from the heat, and set it aside to cool slowly. Set the beaker in an ice bath. Beautiful needle-like crystals of acetylsalicylic acid form.

(iv) Collect the aspirin by vacuum filtration. Wash the crystals with two 10 ml volumes of ice water. Maintain the vacuum to air dry the aspirin. If time does not permit, place the filter paper and aspirin sample on a watch glass and allow them to air-dry. The time for air-drying the sample may require that it be left with your instructor until the next laboratory period.

(v) Transfer the dry aspirin crystals to a pre-weighed sample container or vial. Determine the mass of the aspirin crystals.

DETERMINE THE MELTING POINT OF THE ASPIRIN SAMPLE

Fill a capillary melting point tube to a depth of 0.2 cm with the recrystallized aspirin.

Place the capillary tube in the melting point apparatus. Determine its melting point.

Pure aspirin melts at 135°C.

The aspirin sample should be labeled with your name, the mass of the aspirin, the percent yield, and its melting point.

Note: Don't use your aspirin for a headache! Its purity is not assured.

VERIFICATION OF ASPIRIN

Test for the presence of the phenolic group of the reactant by dissolving a crystal in 1 ml of water and adding 2 drops of $FeCl_3$. As a control, compare any coloration with that produced by a similar quantity of salicylic acid. The experiment may be left overnight after filtering, or during recrystallization.

Submit your product sample in a bottle labelled with: the name of the product, the melting point range, the weight of the product and the percentage yield and your name.

SAFETY PRECAUTIONS

Wear safety glasses or goggles at all times in the laboratory.

Acetic anhydride is corrosive and its vapor is irritating to the respiratory system. Avoid skin contact and inhalation of the vapors. In the event of skin contact, rinse well with cold water. If the vapors are inhaled, move to an area where fresh air is available.

Sulfuric acid is corrosive. Avoid skin contact. In the event of skin contact, rinse well with cold water.

RESULT

Yield obtained = ——— g

EXPERIMENT - 46 Preparation of Benzimidazole

INTRODUCTION

Heterocyclic compounds (Heterocycles) are cyclic compounds with one or more atoms of the ring are atom other than carbon. The name comes from the Greek word *heteros, which means* "different". A variety of atoms such as N , O, S, Se, P and As can be incorporated into the ring structure. Heterocycles are important class of compounds and include many important drugs, and most vitamisns.

Saturated heterocyclic compounds do not contain double bond in its structure. A saturated heterocycle can be named as cycloalkane after prefixing "aza" for nitrogen, "Oxa" for oxygen and "Thia" for sulfur.

Oxacyclopropane

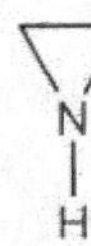

Azacyclopropane

Thiacyclopropane

Pyrrolidine

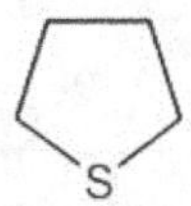

Tetrahydrothiophene

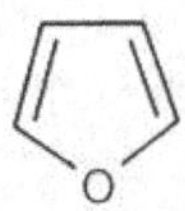

Furan

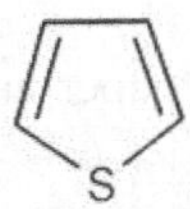

Thiophene

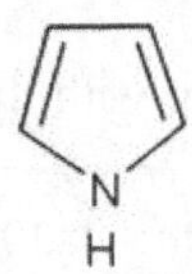

Thiophene

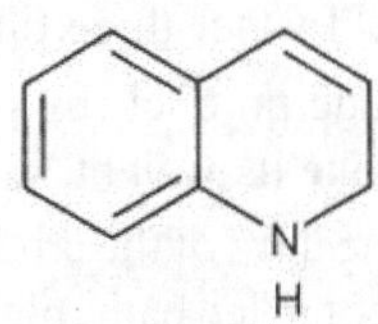

Indole

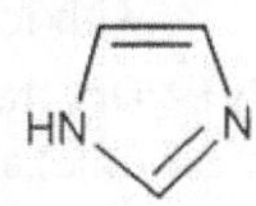

Imidazole

Pyrrole, furan and thiophene are five-membered unsaturated heterocyclic compounds and each has three pairs of delocalized p-electrons and is aromatic.

Benzimidazole contains an imidazole ring fused to a benzene ring. Benzimidazole and its derivatives are used as vermicides or fungicides as they inhibit the action of certain microorganisms. Examples of benzimidazole class fungicides include benomyl, carbendazim, chiorfenazole, cypendazole, debacarb, fuberidazole, furophanate, mecarbinzid, rabenzazole, thiabendazole, thiophanate.

AIM

To prepare benzimidazole using *o*-phenylenediamine.

APPARATUS

1. Conical flask (150 ml)
2. Condenser
3. pH paper
4. Buchner funnel
5. Melting point tube

CHEMICALS

- *o*- Phenylenediamine (OPDA) - 8 grams
- Formic acid (90%) - 5ml
- 10% NaOH solution

PROCEDURE

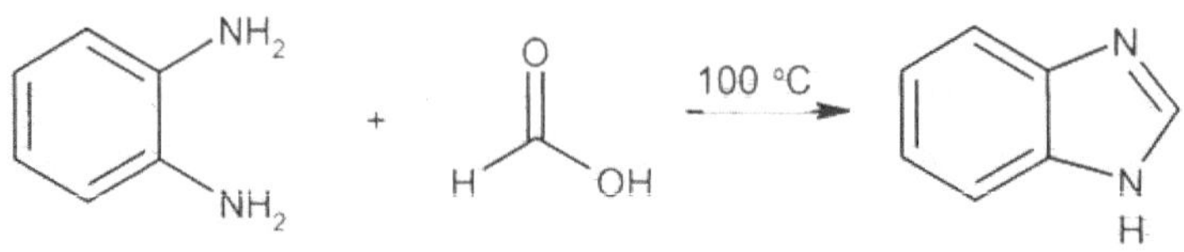

OPDA **Formic acid** **Benzimidazole**

(i) In a 150 mL conical flask fitted with an air condenser, take 8 g of o-phenylenediarnirie and 5 ml formic acid.

(ii) Place the flask in a water bath and heat for 1 hr at 100°C (use this time to take the melting point of o-phenylenediarnine).

(iii) Cool the reaction mixture and adjust the pH to alkaline (monitor the pH of the solution by using pH paper) by drop wise addition of 10% aq. NaOH with constant shaking.

(iv) Filter the precipitate thus obtained by using a Büchner funnel with suction. Wash the precipitate with ice cold water three times.

(v) Dry the crude product (by pressing with the filter paper and then keep it in the oven for some time) and take its weight.

(vi) Dissolve one third of the crude product in 30 ml boiling water. Allow it to cool to room temperature and then cool in ice bath (please do not keep hot beaker in the ice bath), when pure benzimidazole separate out (note 1 and 2).

(vii) Filter it through a fluted filter paper (note 3), dry and take the melting point.

Notes

1. The crude benzimidazole is yellow tinged. This discoloration is difficult to remove and persists after two crystallizations. This discoloration can be removed by treatment with animal charcoal (*not performed here*).
2. The recrystallization begins immediately on cooling. The filtration must be rapid.
3. Fluted filter paper is used to separate a liquid and a solid. This arrangement of folds in the filter paper will allow the liquid to pass through it very quickly and give more surface area on which to collect the crystals.

RESULTS

Total amount of benzimidazole obtained = ——— g

Percentage Yield = ——— %

Melting point of o-phenylenediarnine = ——— °C

Melting point of Benzimidazole = ——— °C

LAB SAFETY INSTRUCTIONS

Lab Safety Guide

Student Responsibilities:

1. Determine the purpose and procedure of the experiment by reading completely the experiment before actually beginning.
2. Wear proper protective equipment.
3. Be aware of the dangers of long hair, long sleeves, and loose clothing.
4. Laboratory groups will be assigned. Remain in your lab group throughout the experiment. Lab activities are team efforts.
5. Do the experiments as assigned and in the manner prescribed. Unauthorized experimentation is not permitted.
6. Keep your lab and other working areas neat and clean during lab sessions.
7. Running, horseplay and practical jokes are NOT allowed. Stay on-task and maintain quiet behavior during lab sessions. Loud and boisterous behavior is not acceptable.
8. Dispose of materials in the proper containers as instructed by your teacher. German law is extremely specific in reference to the disposal chemicals.
9. To avoid poisoning and/or contamination, no eating or drinking is allowed in the science classroom or laboratory.
10. Know the location and proper use of the emergency safety materials such as the fire blanket, eyewash, and fire extinguisher.
11. The teacher MUST be notified IMMEDIATELY of any accident, even if minor.
12. All laboratory equipment MUST be properly cleaned, dried and put away at the end of the lab session. This includes lab aprons, goggles, general lab areas, and sinks.

Chemical Safety

1. Replace caps on chemical bottles immediately after use.
2. NEVER return unused solutions or solids to stock containers as they can contaminate the stock chemical. Report suspected contamination to the teacher.
3. Use paper to protect balance pans. Chemicals SHOULD NOT be placed directly on the balance pan.
4. Never taste chemicals or drink from laboratory glassware.
5. Test the ador of chemicals by waving your hand over the container and sniffing cautiously. Avoid inhaling chemical fumes.
6. Consider ALL chemicals to be dangerous. Most of them can be dangerous if used incorrectly.
7. CLEAN UP IMMEDIATELY ANY CHEMICAL SPILLS. Report any hazardous conditions.
8. Laboratory counters and tabletops WILL be cleaned and dried after each laboratory session.
9. FOLLOW DIRECTIONS for the proper disposal of wastes.

10. *Eye Safety*: If an accident occurs which involves splashing any chemicals into the eye, rinsing or washing of the eye must start immediately and continue for a minimum of 15 minutes. Student will be REQUIRED to wear protective goggles during laboratory periods that can involve any danger to the eyes.
11. *Teacher Instructions:* Any additional safety instruction associated with any experiment as determined by the teacher MUST be followed.

Safety in the Chemical Laboratory

It is very important that you are completely familiar with the experiment before beginning it. Read the experimental details carefully and use the correct safe techniques demonstrated and discussed by the instructor/lab staff. The best precaution against accident is a well-prepared experiment and a neat and well-organized laboratory bench. Planning your experiments in advance not only minimizes accidents but also makes your work proceed more rapidly and smoothly. Whatsoever, **Safety Rules should be followed strictly.** The excuse that someone else is responsible for mistakes/accidents is *never* acceptable in a scientific laboratory. The following regulations are an absolute necessity for a safe laboratory. Please read them carefully. Since there are personal hazard risks associated with non-compliance (fire, explosions, burns, etc.), you may be asked to leave the laboratory immediately if you don't follow these rules.

1. Report any accidents immediately to your instructor or Laboratory staffs.
2. **Do not sit on bench tops.**
3. Contact lenses should never be worn in the laboratory.
4. Be aware of the location of the safety shower, eyewash station, fire extinguisher, first aid equipment, and the exits.
5. Know the hazardous properties of the chemicals you are using. Specific instructions for dealing with hazardous material will be given by the instructor prior to their use. **When in doubt, ask!**
6. Use only the chemicals called for in your experiment. Make sure you know what chemicals you are looking for. Many chemicals have similar names or formulas. Treat all chemicals as hazardous.
7. Flammable liquids and toxic chemicals should be placed specially marked bottles. Paper, broken glass etc should be disposed of in the waste disposal containers provided.
8. All chemicals should be considered potentially toxic and so never taste a chemical or solution or bring eatables into the laboratory. If you are asked to smell a chemical, gently fan the vapors towards your nose. Wash thoroughly your hands with soap after every lab session before touching any foodstuff. If any chemical stain remains on your hand, use spoon to eat till the stain goes off.
9. If you have to insert a glass tube into a rubber or plastic tube, cover the glass tube with a thick towel so as to protect your hand. If you find it difficult to do so, get help from your tutor or lab assistants.
10. Do not keep any organic solvents near a Bunsen burner. If there is a fire of any sort, report to your tutor/instructor immediately and follow his/her directions.

11. Never pipette out using your mouth anything unless otherwise the instructor/tutor has specifically told you to do so. Do not lean on the workbench while working. While using ammonia, remember to keep the exhaust fans on. The same is applicable if any noxious fumes are evolved during a reaction. If you have accidentally taken inside your mouth any solution, wash your mouth immediately with lots of water and contact immediately your instructor or tutor for further help.
12. If any acid or base falls on your hand, immediately wash with lots of water and contact your tutor if there is any uneasiness. While heating solutions over the flame make sure that the mouth of the vessel is not directed towards you or others working nearby.

No "walkmans" and other music devices, no cellular phones with audible ringing tones. (These are strictly forbidden and no exceptions to this rule)

Be fully alert on what you are doing and what is going on around you while working in the laboratory.

In all cases of accident or injury, no matter how slight, report to the instructor or lab staff immediately. While the emphasis should be on the prevention of accidents, a high priority is placed on personal safety in the event of an accident.

Most accidents in the laboratory can be avoided if a few precautions are taken. Since it is difficult to predict when accidents will occur, number of rules must be followed at all times. These rules are listed below:

1. **Read the Label carefully before removing a chemical from its container.**
2. If any of your glass equipment is cracked or broken, do not use it.
3. **If you need to observe the odour of a substance, gently fan a little of the vapour toward you with your hand,** but keep your face at a safe distance.
4. When diluting sulphuric acid, pour the acid slowly into water, with stirring. **Never add water to the acid.**
5. **Never throw sodium metal in the sink,** Sodium metal explodes violently upon contact with water.
6. **Protect your hands if you have to pick up hot objects.** Use tongs whenever practicable, or towels.
7. Use a suction bulb.
8. Use no open flames near flammable solvents. Fire can be avoided by not lighting a burner until you are completely sure that none of you is handling an inflammable material. Ether, benzene, ethanol, and acetone are particularly dangerous. They should be heated or evaporated on a steam bath (hot-water bath), not over a bunsen burner.
9. Before you leave the laboratory, make certain that any gas line you used is shut tightly.
10. If you conducted properly, some chemical reactions can get out of hand and proceed violently. Understand the principles involved in each experiment. **Follow the directions of your teacher and exercise appropriate care.**

Preventive Measures or Habits of Safety

The following precautions, if they are observed sincerely, would avert the most mishaps.

1. Come to laboratory well prepared. For this, read the theory, procedure, and precaution of the experiment to be performed.
2. While working with noxious gases conduct the experiment in a fume hood. Bromine, acetyl chloride, phosphorus trichloride, etc are noxious materials. Therefore they should be carefully handled in the hood. Sodium metal should be used with extra care and its scraps should never be thrown into the sink.
3. Strong oxidizing agents and easily oxidized materials must be mixed with extreme caution and in small amounts. Never add HNO_3 to a flask containing alcohol or other easily oxidized material. The reaction between HNO_3 and organic reducing can be so violent that a dangerous explosion may result.

First Aid Box

A first aid box or cupboard should be kept in a readily accessible position in the laboratory and should contain the following items, clearly labeled:

1. Bandages (several sizes), gauze, lint, cotton wool, adhesive plaster, wound dressings (various sizes), and a sling (triangular bandage).
2. Delicate forceps, scissors, and safety pins.
3. Eye bath, empty wash bottle, eye dropper.
4. Table salt, sodium hydrogen sulphate powder, glycerol, paracetamol tablet, pentyl nitrite capsules.
5. 0.5% Cretimide cream, *e.g.*, Savlon
6. Burnol, Furacin, or Acriflavin.
7. Bottles containing:
   (a) 1% acetic acid (the bottle should be marked: Not for the eyes) solution.
   (b) 1% boric acid (H_3BO_3) solution.
   (c) 1% sodium hydrogen carbonate solution.
8. Milk of magnesia
9. A reliable disinfectant, *e.g.*, dettol.

First Aid Measures for Laboratory Accidents

In case of a mishap, never lose your presence of mind and immediately inform the teacher. If the case is a serious one, call for a doctor immediately. The following first aid measures should be given before seeking medical attention:

Accidents	First Aid Treatment
1. Cuts and Wounds	1. Remove any foreign bodies, e.g., glass pieces, etc., which are visible. Clean the wound with a sterile gauze. 2. Stop the bleeding. Remember the following three P's (*Position, pressure, and Packing*) for this purpose: (a) *Position:* Raise up the injured organ:

Accidents	First Aid Treatment
	(b) *Pressure:* Apply gentle pressure with the thumbs or fingers at the place just before injury. Never apply a very strong pressure for more than 5 minutes.
	(c) *Packing:* Apply cotton soaked in alum or ferric chloride solution. If an internal organ is injured such as in a nostril, it should be packed with this cotton.
	3. Wash with soap and water. Disinfect with an antiseptic solution (Spirit, Dettol, Savlon or 2% Iodine solution), spray Cebazole or Sulphona-midpowder and bandage with Leucoplast.
2. Burns and Scalds	Put the affected area immediately in water, preferably ice cold. Avoid handling affected area as far as possible.
(i) Acid burn	Wash immediately with large quantities of water, then with a very dilute (1%) solution of sodium bicarbonate and apply an oil or a cream dressing as above. (If the burn is by H_2SO_4, do not wash with water but only with $NaHCO_3$ solution). Caution: If the concentration of $NaHCO_3$ is more than 8% it will give burning pain instead of relief.
(ii) Caustic alkali burn	Wash immediately with large quantities of water, then with dilute (preferably 1%) acetic acid or lemon juice and apply an oil or a cream dressing.
(iii) Sodium burn	Remove any piece of Na with the help of forceps. Wash with a plenty of water and then with dilute acetic acid. Apply an oil or a cream bandage.
3. Eye Injuries	Remove the solid (glass piece, chemical particle, insect, etc.)
(i) By an acid or an alkali	Wash with large quantities of water and then with 1% $NaHCO_3$ solution *in case of an acid and with* 1% boric *in case of an alkali.* Cover the eye with gauze moistened with olive or castor oil.
4. Fires: (i) Inflammable substance fire	(a) Immediately cover the flame or burning substance so that the oxygen supply is cut off. For example, a spirit lamp on fire can be covered with an empty beaker. A small beaker should be covered with a big one.
	(b) If the fire has spread on the table or floor, turn of all the burners and gas pipes and cover it with the sand from sand-batch or soil from any where. If Na, K, petrol, oil or an organic solvent is not burning, pour water carefully.
	(c) If Na, K, gasoline, oil or an organic solvent is on fire, only *dry* sand should be used. Never use water in this case or the fire will become more violent.

Accidents	First Aid Treatment
(ii) Burning of clothes	Cover the burning clothes with a blanket, towel, or a piece of thick cloth. If such a cloth is not available, ask the victim to roll on the floor and pour water.
5. Inhalation of poisonous fumes and gases	Remove the victim to the open air and loosen clothing at the neck. If breathing has stopped, give artificial respiration until help arrives.

NOTES ON LABORATORY FIRST AID

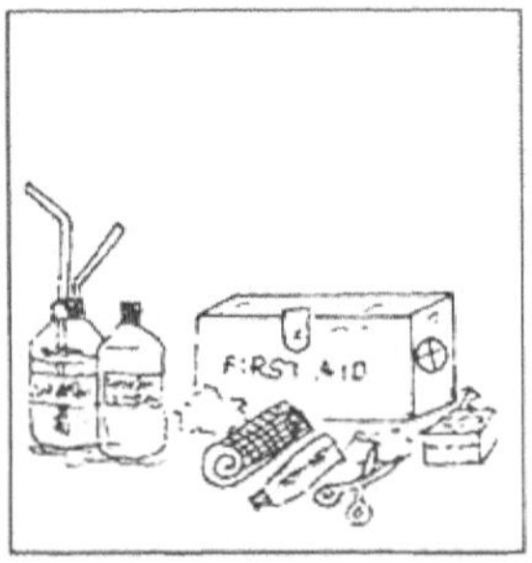

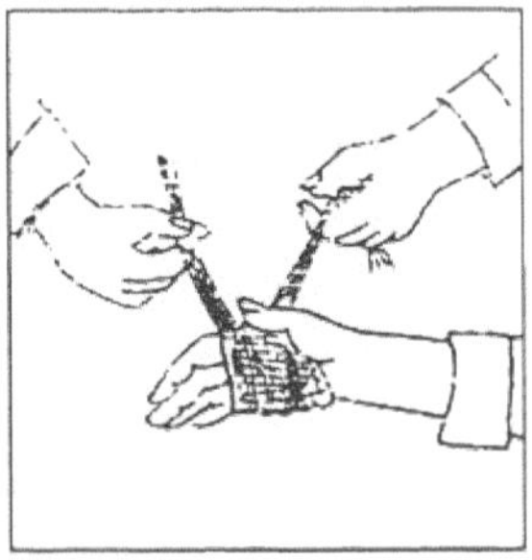

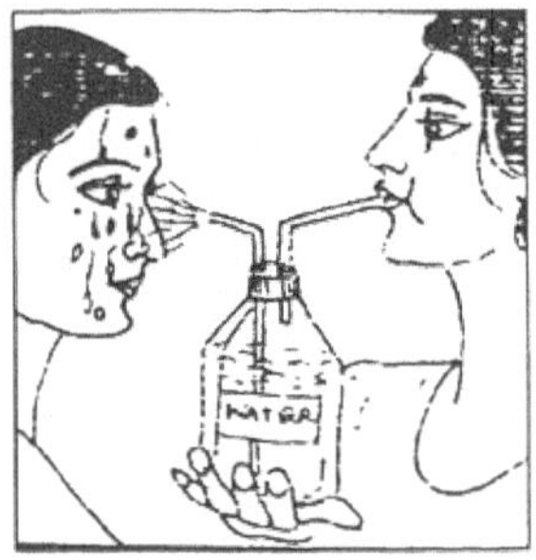

TREATMENT OF CUTS AND SCRATCHES

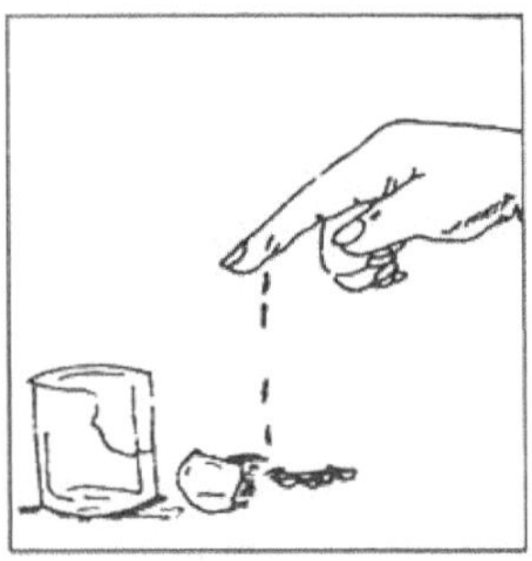

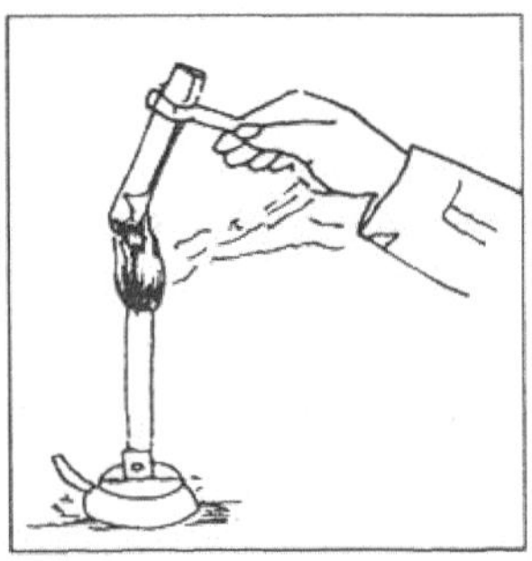

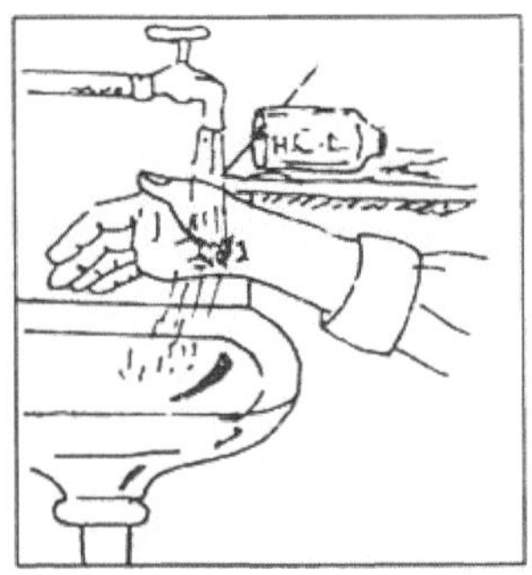

	CHEMICAL HEZARD	AFFECTED PART OF BODY	FIRST AID
1.	ACETIC ACID	Lungs	Remove from exposure: rest and keep warm.
		Skin	Drench the skin with plenty of water, remove contaminated clothing and wash before reuse, in severe cases, OBTAIN MEDICAL ATTENTION.
		Mouth	Wash out the mouth thoroughly with water and give plenty of water to drink, followed by milk of magnesia. OBTAIN MEDICAL ATTENTION.

Contd.....

	CHEMICAL HEZARD	AFFECTED PART OF BODY	FIRST AID
		Lungs	Remove from exposure, rest and keep warm: in severe cases or if exposure has been great, OBTAIN MEDICAL ATTENTION.
2.	AMMONIUM HYDROXIDE	Skin	Drench the skin with plenty of water, remove contaminated clothing and wash before reuse, in severe cases, OBTAIN MEDICAL ATTENTION.
		Mouth	Wash out the mouth thoroughly with water, give plenty of water, followed by vinegar or 1 % acetic acid to drink. OBTAIN MEDICAL ATTENTION.
3.	CAUSTIC POTASH	Skin	Drench the skin with plenty of water, remove contaminated clothing and wash before reuse in severe cases. OBTAIN MEDICAL ATTENTION.
		Mouth	Wash out the mouth thoroughly with water: give plenty of water followed by vinegar or 1 % acetic acid to drink. OBTAIN MEDICAL ATTENTION.
4.	CAUSTIC SODA	Skin	Drench the skin with plenty of water, remove contaminated clothing and wash before reuse, in severe cases. OBTAIN MEDICAL ATTENTION.
		Mouth	Wash out the mouth thoroughly with water, give plenty of water, followed by vinegar or 1 % acetic acid to drink. OBTAIN MEDICAL ATTENTION.
5.	CHLORINE	Lungs	Remove from exposure, rest and keep warm. In severe cases or if exposure has been great. OBTAIN MEDICAL ATTENTION.
		Skin	Drench the skin with plenty of water, remove contaminated clothing and wash before reuse, in severe cases. OBTAIN MEDICAL ATTENTION.
6.	CHLOROFORM	Lungs	Remove from exposure, rest and keep warm: in severe cases. OBTAIN MEDICAL ATTENTION and apply artificial respiration if breathing has stopped.
		Mouth	Wash out the mouth thoroughly with water and give an emetic. OBTAIN MEDICAL ATTENTION.
7.	CHROMIC ACID	Lungs	Remove from exposure, rest and keep warm.
		Skin	Drench the skin with plenty of water, remove contaminated clothing and wash before reuse, in severe cases. OBTAIN MEDICAL ATTENTION.
		Mouth	Wash out the mouth thoroughly with water and give plenty of water to drink, followed by milk of magnesia. OBTAIN MEDICAL ATTENTION.

Contd.....

	CHEMICAL HEZARD	AFFECTED PART OF BODY	FIRST AID
8.	ETHER	Lungs	Remove from exposure, rest and keep warm. In severe cases or if exposure has been great. OBTAIN MEDICAL ATTENTION.
		Mouth	Wash out the mouth thoroughly with water and give an emetic. OBTAIN MEDICAL ATTENTION.
9.	HYDROCHLORIC ACID	Lungs	Remove from exposure, rest and keep warm.
		Skin	Drench the skin with plenty of water, remove contaminated clothing and wash before reuse, in severe cases, OBTAIN MEDICAL ATTENTION.
		Mouth	Wash out the mouth thoroughly with water and give plenty of water to drink, followed by milk of magnesia, OBTAIN MEDICAL ATTENTION.
10.	HYDROGEN PEROXIDE	Skin	Drench the skin with plenty of water, remove contaminated clothing and wash before reuse, in severe cases, OBTAIN MEDICAL ATTENTION.
		Mouth	Wash out the mouth thoroughly with water and give large quantities of water to drink. OBTAIN MEDICAL ATTENTION.
11.	HYDROGEN SULPHIDE	Lungs	Remove from exposure, rest and keep warm. In severe cases OBTAIN MEDICAL ATTENTION and apply artificial respiration if breathing has stopped.
12.	METHANOL	Lungs	Remove from exposure, rest and keep warm. In severe cases if exposure has been great. OBTAIN MEDICAL ATTENTION.
		Skin	Drench the skin with water and wash with soap and water. Remove contaminated clothing and wash before reuse (Ga: clothing to be thoroughly aired instead of washed).
		Mouth	Wash out the mouth thoroughly with water and give an emetic. OBTAIN MEDICAL ATTENTION.
13.	NITRIC ACID	Lungs	Remove from exposure, rest and keep warm. In severe cases or if exposure has been great. OBTAIN MEDICAL ATTENTION.
		Skin	Drench the skin with plenty of water, remove contaminated clothing and wash before reuse, in severe cases, OBTAIN MEDICAL ATTENTION.
		Mouth	Wash out the mouth thoroughly with water and give plenty of water to drink, followed by milk of magnesia, OBTAIN MEDICAL ATTENTION.

Contd.....

	CHEMICAL HEZARD	AFFECTED PART OF BODY	FIRST AID
14.	ORTHOPHOSPHORIC ACID	Skin	Drench the skin with plenty of water. remove contaminated clothing and wash before reuse in severe cases. OBTAIN MEDICAL ATTENTION.
		Mouth	Wash out the mouth thoroughly with water and give plenty of water to drink. followed by milk of magnesia. OBTAIN MEDICAL ATTENTION.
15.	PHOSPHORIC ACID	Skin	Drench the skin with plenty of water. Remove contaminated clothing and wash before reuse. In severe cases. OBTAIN MEDICAL ATTENTION.
		Mouth	Wash out the mouth thoroughly with water and give plenty water to drink followed by milk of magnesia. OBTAIN MEDICAL ATTENTION.
16.	POTASSIUM CYANIDE	Lungs	Remove from exposure rest and keep warm. If breathing. break a capsule of amyl nitrite and give to casualty by inhalation for 15–30 seconds. Repeat every 2–3 minutes. Apply artificial respiration if breathing has stopped. In any case. OBTAIN MEDICAL ATTENTION at once.
		Skin	Drench the skin with plenty of water. Remove contaminated clothing and wash before reuse. in severe cases. OBTAIN MEDICAL ATTENTION.
		Mouth	Give cyanide antidote if breathing. break capsule of amyl nitrate and give to inhale for 15–30 seconds. repeat every 2–3 minutes. Apply artificial respiration if breathing has stopped. In any case. OBTAIN MEDICAL ATTENTION.
17.	SODIUM HYPOCHLORITE SOLUTION	Skin	Drench the skin with plenty of water. remove contaminated clothing and wash before reuse. in severe cases. OBTAIN MEDICAL ATTENTION.
18.	SULPHUR DIOXIDE	Lungs	Remove from exposure. rest and keep warm. In severe cases or if exposure has been great. OBTAIN MEDICAL ATTENTION.
19.	SULPHURIC ACID	Skin	Drench the skin with water BUSTERS OR BURNS MUST RECEIVE MEDICAL ATTENTION. Remove contaminated clothing and wash before reuse.
		Mouth	Wash out the mouth thoroughly with water and give plenty of water to drink. followed by milk of magnesia. OBTAIN MEDICAL ATTENTION.

SOME DON'TS IN THE LABORATORY

1. Don't handle the apparatus roughly. it may result in damage and breakage.
2. Don't perform the experiment with incomplete knowledge; it may lead you in danger.
3. Don't consult your fellow students. if you have a doubt about the experiment. He may mislead you. Consult the lecturer incharge for the laboratory.
4. Don't put on inflammable clothes and don't enter into the lab without proper footwear.

5. Don't use excess amount of chemicals or reagents. It may spoil your results. It may also contaminate the laboratory with excess of fumes or gases.
6. Don't keep the reagent bottles on your work tables after completing their use. It may lead to your confusion.
7. Don't forget to stopper the reagent bottles immediately after use.
8. Don't handle used glass rods or spatulae without cleaning otherwise they may contaminate your process.
9. Don't heat a substance by keeping the test tube near your eyes or near the water tap.
10. Don't throw the hot solutions directly into the sink. Allow them to cool and clean the test tube.
11. Don't throw broken pieces of apparatus, splinters and used up filter papers into the sink. Throw them in a separate porcelain or glass trough provided for this purpose.
12. In a chemical laboratory don't taste any chemical. Most of the chemicals are dangerous and poisonous.
13. Don't waste water and gas when they are not required: the water and gas taps must be closed.
14. Don't inhale poisonous and suffocating vapours. Conduct the reactions (involving the production of such vapours)using minimum quantities of reagents. After the tests are over, immediately clean the test tubes.
15. Don't use laboratory water for drinking purpose.
16. Don't waste any reagent or solvent or solution. Sometimes excess of reagents leads to adverse results.
17. Don't throw any solid matter in the sink.
18. Don't throw conc. acids into sink. Be careful with hot conc. H_2SO_4. Cool and dilute the acid with water, then drop it slowly into the sink.
19. Don't take the reagents from the bottles with your naked fingers.
20. Don't return unused chemicals to the stock bottles.
21. Don't write your observations on loose sheets. Record them in the observation notebook only.
22. Don't enter into the laboratory without the following articles: 1. Chemistry practical manual, 2. Observation notebook, 3. An apron, 4. A clean handkerchief and 5. Shoes on your legs.
23. Don't exchange the stoppers of one bottle to another.
24. Don't leave the laboratory unless your workbench is clean and all the apparatus is returned to the attender.
25. Don't dip glass rods or filter paper strips in reagent bottles.

GENERAL INTRODUCTION

How to use a burette: The burette should be washed thoroughly before using. When mounting the burette, make sure that it is clamped vertically and the tap is on your right side. It is then filled up to the zero mark and the nozzle is filled by opening the tap for two or three seconds. The reading of the bottom level of the meniscus is taken. The tap is held by left hand (Thump and fore-finger). The flask containing the solution is held in the right hand. During the titration, the solution is added slowly from the burette and the flask is whirled with right hand. The burette after use should be emptied and washed thoroughly with water before leaving the lab.

How to use a pipette: The pipette should be thoroughly washed before using. Hold the pipette in the right hand and keep your fore-finger free. Suck the liquid. till the liquid level reaches a little above the mark. Take it from your mouth and close the top with your fore finger. The pipette is then raised at your eye-level and by controlled release of the fore finger; the liquid level is adjusted to the mark. The lower end of the pipette is then introduced into a conical flask and by removing the fore-finger. the liquid can be drained off. A little liquid will remain there in the tip and no attempt should be made to remove because the required volume has already delivered in the flask.

How to use a standard measuring flask: Standard flask is a narrow necked flask fitted with a glass stopper. A weighed amount of the substance is usually transferred to it by using a funnel. The funnel and its stems are washed down into the flask. The funnel is removed and the substance is then dissolved by giving a rotating motion to the flask. Water is then added to it till the lower water meniscus coincides with the mark on the standard flask. The flask is then closed and shaken well till the solution gets a uniform concentration.

End Point: The end point of a titration is the point at which complete reaction takes place between two solutions. The end point is usually determined by using an indicator which shows a marked color change at the point of completion of the reaction.

Indicator	pH range	Color in acid solution	Color in alkaline solution
Methyl orange	3.1 to 4.4	Red	Orange
Phenolphthalein	8.3 to 10	Colorless	Pink

GENERAL INSTRUCTIONS

1. All apparatus should be extremely clean to start with. Test tubes. beakers can be cleaned with soap powder with a semimicro brush. At the end of the day's work. you must clean all the apparatus and return to the lab staff.
2. Droppers containing the reagents should not be allowed to touch on the sides of the test tube during addition. Reagent drops should be allowed to fall directly into the test tube without touching the sides; otherwise the reagent bottle will get contaminated.

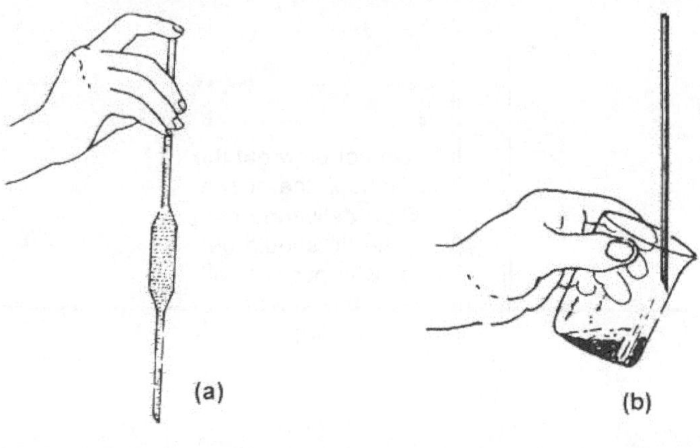

(a) (b)

Transferring of a liquid with a pipette

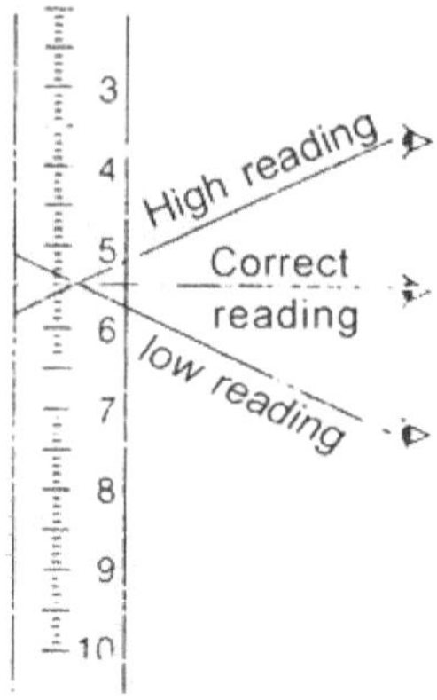

Taking the burette reading.

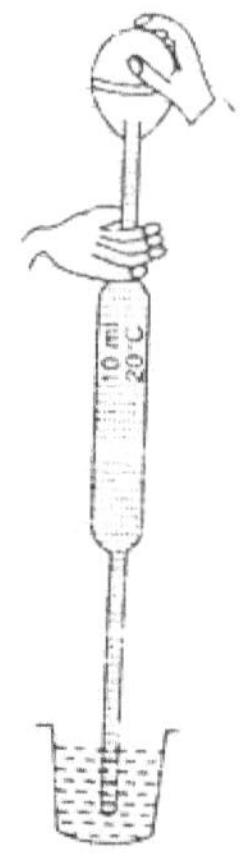

Suck the liquid in the pipette using rubber suction bulb

Do not blow out the liquid in the nozzle When delivering stops, the tip should be slowly touched with the wall of flask

Never blow out the last drop clinging on the nozzle because it has already been accounted for while calibrating

Graduated Pipette consists of straight fairly narrow tube without any central bulb. There are three main types:

- Those which deliver a measured volume from a top zero to a selected graduation.
- Those which deliver a measured volume from a selected graduation mark to the nozzle (zero is at the nozzle)
- Those calibrated to contain a given capacity from the nozzle to a selected mask.

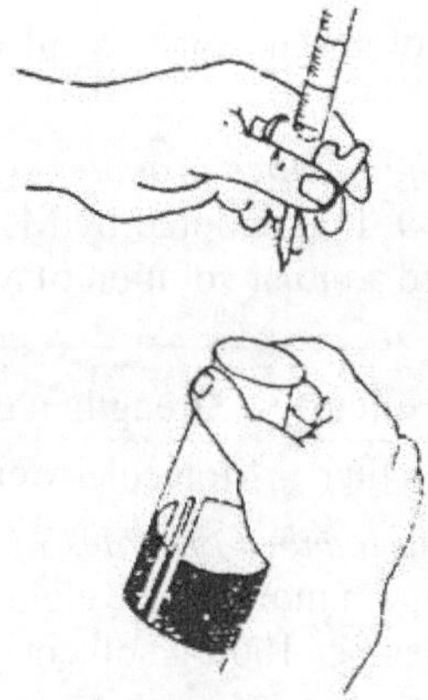

Titration

I WASH BOTTLE

You require a wash bottle to wash with distilled water the glass apparatus used in chemical analysis. to wash the precipitates in quantitative analysis and to add distilled water carefully to solutions in the volumetric flasks in the process of making up. A wash bottle is a flat bottomed flask fitted with two glass tubes as shown in Fig. 7. The flask is made up of glass or polythene.

II FILTERING FUNNEL

This is the simplest apparatus used for filtration with the help of a filter paper placed in it. The funnel should have an angle as close to 60° as possible. A long stem is preferred to promote rapid Filtration. Filter papers for quantitative work are made in varying grades of porosity. The filtering funnels too are available in different capacities.

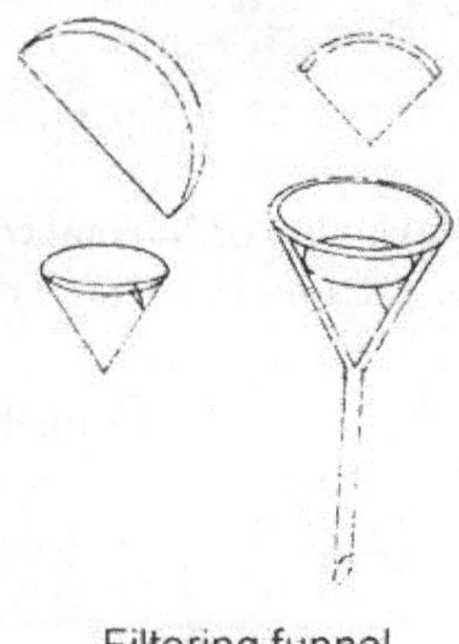

Filtering funnel

VOLUMETRIC ANALYSIS

Volumetric analysis or volumetry is the branch of analytical chemistry in which measurement by volumes is the main and final operation.

CONCENTRATION OR STRENGTH OF A SOLUTION

Concentration or Strength of a solution is *the mass of a solute dissolved in the unit volume of a solution. It is expressed in several ways.* The most common are:

1. **Grams per unit volume.** *e.g.*, g/l. g/ml. etc. If 5 g of solute is dissolved in 250 ml of solution. the strength is 20 g/l.

2. **Parts per million (ppm),** *e.g.*, mg/l, g/million ml, etc. A solution containing 10 milligrams of solute per liter or 10 micrograms per milliliter of solution is a 10 ppm solution.

$$1\ \mu g = 0.000001\ g$$

3. **Percentage** consists of expressing strength in terms of grams or ml of solute per 100 g or 100 ml of solution.
   (i) *By weight:* A 5% solution by weight of a given salt means 5 g of salt is dissolved in 95 g of solvent.
   (ii) *By volume:* A 25% solution by volume means 25 ml of solute is dissolved in 75 ml of solvent.

4. **Molarity of Molar concentration** is *the number of moles* (gram molecular weights) *of solute dissolved in a liter* (1000 ml) *of solution.* It is denoted by M. When a gram molecular weight is dissolved in a liter of solution. It is called a molar solution of M solute. Similarly 2 moles per liter is 2M solution.

$$M = \frac{\text{Number of moles of solute}}{\text{Volume of solution in liter}} = \frac{\text{Strength in g/l}}{\text{Molecular weight}}$$

5. **Molality or Molal concentration** is *the number of moles of solute dissolved in* 1000 *grams of solvent.* A molal solution contains one gram molecular weight of substance, dissolved in 100 g of solvent. It is denoted by *m*. Thus 3 moles per 1000 g solvents is a 2 *m* solution.

$$\text{Molality } (m) = \frac{\text{Number of moles of solute}}{\text{Weight of solvent per 1000 g or Number of kg's of solvent}}$$

$$= \frac{\text{Moles of solute}}{\text{Kilograms of solvent}}$$

6. **Formality** or **Formal concentration** is *the number of formula weights dissolved in one liter of solution.* It is denoted by F. For ions such as SO_4^{2-}, $Cr_2O_7^{2-}$, etc., the term *formula weights* is more appropriate than *molecular weight* because they are not molecules. Hence it is formality of SO_4^{2-}, $Cr_2O_7^{2-}$ which is more correct than molarity.

$$\text{Formality} = \frac{\text{Strength in g}}{\text{Formula weight}}$$

7. **Normality** or **Normal concentration** *is the number of gram equivalent weights of solute dissolved in one liter of solution or number of gram milliequivalents of solute per milliliter of solution.*

$$\text{Normality} = \frac{\text{Strength in g/l}}{\text{Equivalence weight}}$$

$$= \frac{\text{Number of gram equivalence of solute}}{\text{Number of liters of solution or volume of solution in liter}}$$

$$= \frac{\text{Number of gram equivalents of solute}}{\text{Number of milliliters of solution}}$$

$$= \text{Molarity} \times \text{Acidity, Basicity or Change in oxidation number per atom.}$$

8. **Mole fraction** of a solute is *the ratio of the number of its moles to the total number of moles of both the solute and the solvent:*

$$\text{Mole fraction} = \frac{\text{Number of moles of solute}}{\text{Number of moles of solute + Number of moles of the solvent}}$$

Solutions in redox titrations should be acidic (1-2N) for the required oxidation potential to be achieved.

$$\text{Equivalent weight of an acid, base or a salt} = \frac{\text{Molecular weight}}{\text{Total positive valency}}$$

$$\text{Equivalent weight of an acid} = \frac{\text{Molecular weight of acid}}{\text{Basicity of the acid}}$$

$$\text{Equivalent weight of a base} = \frac{\text{Molecular weight of the base}}{\text{Acidity of the base}}$$

$$\text{Equivalent weight of an oxidant} = \frac{\text{Molecular weight of oxidant}}{\text{Number of electrons gained by one mole of reductant}}$$

$$\text{Equivalent weight of a reductant} = \frac{\text{Molecular weight of reductant}}{\text{Number of electrons lost by one mole of reductant}}$$

$$\text{Equivalent weight} = \frac{\text{Molecular weight}}{\text{Change in oxdation number per mole}}$$

Calculations

For monobasic acids, monoacidic bases and equivalent salts (*i.e.*, salts having ions of same valency) or where there is transference of one electron in a redox system, normality is equal to molarity (N = M) and $N_1V_1 = N_2V_2$ is equal to $M_1V_1 = M_2V_2$.

The general formula for the use of molarity in calculation is:

$$\frac{M_1V_1}{n_1} = \frac{M_2V_2}{n_2} \quad \text{or} \quad \frac{M_1V_1}{M_2V_2} = \frac{n_1}{n_2}$$

where M = molarity, n = number of moles and V = volume of the reactants.

Note that $M_1/n_1 = N_1$ and $M_2/n_2 = N_2$. If $n_1 = 1$ and $n_2 = 1$, $N_1V_1 = N_2V_2$ is equal to $M_1V_1 = M_2V_2$ as in the above case.

For dibasics, diacids or divalents, Molecular weight = 2 × Equivalent weight

$\therefore$ Normality = 2 × Molarity, i.e., N = 2M

$$\therefore \quad M = \frac{N}{2}$$

$\therefore$ Strength = N × Equivalent weight = M × Molecular weight

$$= \frac{N}{2} \times 2 \text{ Equivalent weight}$$

For example, the molecular weight of Na_2CO_3 is 106. If 5.2 g of it is dissolved in 1 liter, its strength = N × Equivalent weight = M × Molecular weight

$$= \frac{5.2}{53} \times 53 = \frac{5.2}{106} \times 106 = 5.2 \text{ g/l}$$

(Equivalent weight of Na_2CO_3) = 106 ÷ 5.2)

For the titration between $KMnO_4$ and ferrous ammonium sulphate (FAS) the ionic reaction is

$$MnO^-_4 + 8H^+ + 5Fe^{2+} \rightarrow Mn^{2+} + 4H_2O + 5Fe^{2+}$$

$$\therefore \quad \frac{M_1V_1 \text{ of } KMnO_4}{M_2V_2 \text{ of FAS}} = \frac{n_1 \text{ of } MnO_4}{n_2 \text{ of } Fe^{2+}} = \frac{1}{5}$$

Indicators

The indicators employed in oxidation-reduction titrations can be of the following types.

(i) ***Internal oxidation-reduction indicator***

They exist in two different colours in their oxidised and reduced forms and undergo the redox reaction only after the titration reaction is completed. This can happen only if their reduction potential is close to the reduction potential of the reacting reagents at the equivalence point. This would result in a sharp change in colour and would hence minimise the titration errors.

$$\underset{\text{(oxidised form)}}{\text{Indicator} + ne^-} \rightleftharpoons \underset{\text{(reduced form)}}{\text{Indicator}}$$

$$E = E^o - \frac{0.059}{n} \log \frac{[\text{In}](\text{reduced form})}{[\text{In}^{n+}](\text{oxidised form})}$$

Some examples of these indicators are – diphenylamine, ferroin, *N*-phenyl anthranillic acid etc.

Diphenylamine is a colourless molecule in the reduced form. On oxidation by $K_2Cr_2O_7$, it gives the deep violet oxidised form

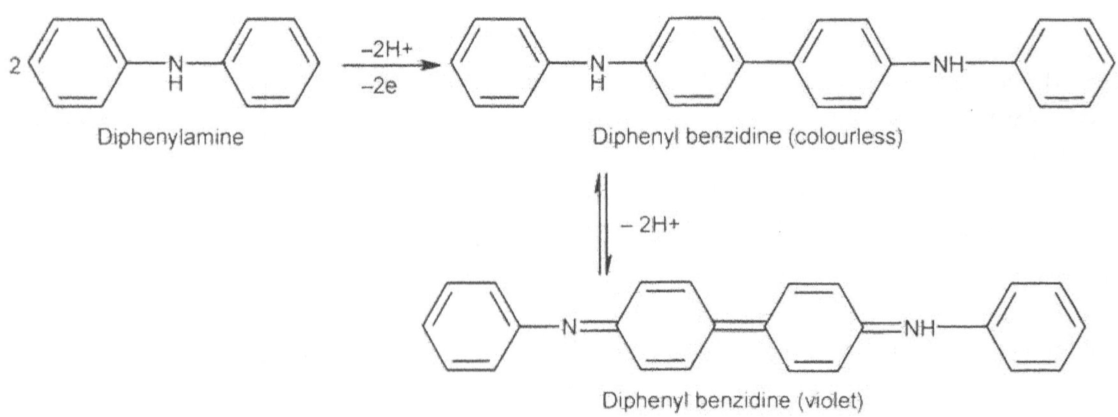

Since diphenyl amine is sparingly soluble in water, usually a sodium salt of its sulphonic acid is used as an indicator.

(ii) ***Self indicator:*** In the titration using $KMnO_4$ as a titrating agent, the reduced form (Mn^{2+}) forms an almost colourless solution. So an long as there is some reducing agent in the reaction mixture the solution remains colourless. But at the end of the titration, all the reducing agent has been consumed. The next drop of $KMnO_4$ is in excess and remains unreduced. Thus the solution turns pink and indicates the end of the titration by changing the colour of the solution from colourless to pink.

(iii) ***Starch as an indicator:*** Iodine complexes with starch to form a deep blue coloured complex. So the solution remains colourless till iodine is consumed. The first drop of excess iodine will change the solution to deep blue. If the iodine is in the reaction mixture then starch is added towards the end of the titration (when the colour has become pale yellow) and the end point is indicated by a colour change from blue to colourless.

(iv) ***External indicator:*** These are used when no internal indicator is available. These indicators are not added to the reaction medium because of two difficulties:

(i) in the case of dark coloured liquids, a sharp change in colour at the end point is not visible:

(ii) when the indicator forms an insoluble precipitate with an ion present in the reaction mixture solution to which it is added.

An example of an external indicator is potassium ferricyanide $K_3[Fe(CN)_6]$ in the titration of $K_2Cr_2O_7$ and $FeSO_4 (NH_4)_2SO_4.6H_2O$ in an acidic medium. The reaction is

$$2K_3[Fe(CN)_6] + 3FeSO_4 \rightarrow Fe_3[Fe(CN)_6]_2 + 3K_2SO_4$$

Ferro-ferric cyanide (deep blue colour)

Types of Volumetric Analysis

Following are some major types of titrations performed in Volumetric Anaiyss:

I. Acid-alkali, Acid-Base or Neutralization Titrations

These are analyzes in which the two solutions used are acid and alkali respectively. These reactions involve neutralization, *i.e.*, the combination of hydrogen and hydroxyl ions to form water.

$$H^+ + OH^- \rightarrow H_2O$$

The process of determining the strength of unknown acid solution by titrating with a standard alkali solution is called *acidimetry*. Similarly, the titration of a free base or alkali with a standard acid is called *alkalimetry*. *Acid-base* or *p*H indicators are either weak organic acids or weak organic bases. They possess different colors in dissociated and undissociated forms and also in different *p*H conditions. Phenolphthalein and methyl orange are the most commonly used *p*H indicators.

II. Oxdition-Reduction or Redox Titrations

These are titrations in which the two solutions used are the solutions of oxidizing the reducing agents respecticely. Thus in a redox reaction, transference of electrons takes place during titration, *i.e.*, one substance is oxidized and the other is reduced. In such titrations either a *reducing agent* (*reductant*) is titrated with a standard *oxidizing agent* (*oxidant*) solution or an oxidant is titrated with a standard reductant. The important oxidants are $KMnO_4$, $K_2Cr_2O_7$, $Ce(SO_4)_2$, I_2, K_2BrO_3, and K_2IO_3; while the common reductants are $FeSO_4$, $(NH_4)_2SO_4 6H_2O$ (Mohr's salt), $SnCl_2$, $Na_2S_2O_3$, As_2O_3, and $TiCl_3$.

The indicators used in redox titrations are called redox indicators. They are mainly of two types:

(i) Specific redox indicators: They react specifically with oxidizing or reducing agents, *e.g.*, starch, thiocyanate, etc. Starch gives a dark-blue colour with iodine, an oxidant. Thiocyanate gives a blood-red color with ferric ion, a reductant.

(ii) True redox indicators: They exhibit different colors in the oxidized and reduced forms. Therefore, they mark sudden changes in the oxidation potential near the equivalence point in a redox titration, just as acid-base indicators mark sudden changes in *p*H during acid-base titrations. Ferroin, diphenylamine, methylene blue, potassium ferricyanide, etc., are examples of some redox indicators.

III Iodine Titrations (Iodimetry and Iodometry)

They are also redox titrations. They are based upon the use of iodine as an oxidant. They are studied separately for the sake of simplicity. Volumetric analysis involving the use of iodine is of two types:

(i) Direct or iodimetric methods involve the use of a standard iodine solution to titrate easily oxidizable substances, *e.g.*, the titration of As_2O_3 and I_2.

$$AS_2O_3 + 2I_2 + 2H_2O \longrightarrow As_2O_5 + 4HI$$

They are of limited application because I_2 is a weaker oxidant than MnO_4, $Cr_2O_7^{2-}$ and Ce^{4+} salts.

(ii) Indirect or iodometric techniques are those titrations in which I_2 liberated from in iodide (say KI) is titrated against standard hypo or sodium thiosulphate ($Na_2S_2O_3$) solution, *e.g.*, the titration of $CuSO_4$ and hypo, of $K_2Cr_2O_7$, and hypo, of $KMnO_4$ and hypo, etc.

$$2Cu^{2+} + 2I^- \rightarrow 2Cu^+ + I_2 \uparrow$$

$$Cr_2O_7^{2-} + 14H^+ + 6I^- \rightarrow 2Cr^{3+} + 7H_2O + 3I_2 \uparrow$$

$$2MnO_4 + 16H^+ + 10I^- \rightarrow 2Mn^{2+} + 8H_2O + 5I_2 \uparrow$$

$$2S_2O_3^{2-} + I_2 \rightarrow S_4O_6^{2-} + 2I^-$$

Iodine solution can act as a self-indicator. However, to get a better result a 1% starch solution is used as an indicator with which even traces of I_2 can give a deep blue color due to the formation of an adsorption complex called *starch iodide*, $(C_{24}H_{40}O_{20})I_3$.

IV. Precipitation Titrations or Precipitimetry

Precipitimetry involves the combination of cations and anions (other than H^+ and OH^-) so as to produce a precipitate or turbidity due to formation of an insoluble salt and the strength of unknown solution is determined by complete precipitation with the help of standard solution, *e.g.*, the titration of $AgNO_3$ and KI, Na_2CrO_4, KCNS, or NaCl where AgI (yellow), Ag_2CrO_4 (bright red), AgCNS (white) or AgCl (white) precipitates, respetively, are formed. Titrations involving $AgNO_3$ are called *argentometric titration*.

V. Complexometric Titration or Compleximetry

These are the titrations involving the formation of a complex ion. Only a few reactions of complexation can be used as a basis for volumetry, partly due to the lack of a suitable indicator and partly because in some cases varying mixtures of products are obtained. Most such methods use standard solutions of silver salts (*argentometry*) and standard solutions of mercury salts (*mercurymetry*). A large number of estimations have been carried out by using complexing agents mostly ethylenediamine-tetraacetic acid (EDTA). The equivalence point is detected by the formation or disappearance of a solid phase, acid-base indicators, or formation or disappearance of a soluble complex.

Special Reangents used in the Organic and Other Analysis

Reagent	Methods of Preparing Solution
Alcohol 1 : 2 [C_2H_5OH]	Add equal amount of water to rectified spirit. The commercial alcohol usually contains 95% C_2H_5OH.
Benedict's solution or Cu^{2+} solution in the presence of citrate ions (For the qualitative test of glucose	Dissolve 17.3 g of $CuSo_4$. $5H_2O$ in 100 ml of water by heating, in another beaker dissolve 17.3 g of sodium citrate and 100 g of anhydrous sodium carbonate in 800 ml of water. Add $CuSO_4$, $5H_2O$ solution to it with constant stirring and make up the volume to 1 liter
Brominating mixture	Dissolve 5.567 g of potassium bromate and 75 g of KBr in water and dilute to 1 liter.
Bromine water	
(*i*) Ordinary	Shake about 5 ml of liquid bromine with 100 ml of water until no more bromine dissolves in it. It is a saturated solution. It spoils after a week.
(*ii*) Strong	Dissolve 1 ml of bromine and 1.5 g of KBr in 10 ml of water.
CCl_4 in bromine	Dissolve 5 ml in 100 ml of CCl_4.
Bleaching powder	Shake 12.5 g of bleaching powder with 100 ml water for an hour. Filter if necessary.
Caustic potash (ethanolic or alcoholic KOH)	Dissolve 11.2 g of caustic potash (KOH) in 100 ml of rectified spirit or ethanol.
Caustic potash (aqueous KOH)	Dissolve 312 g of KOH in water and dilute to 1 liter. This gives 5N KOH.
Caustic soda (alcoholic)	Dissolve 20 g of NaOH in ethanol and dilute with water to 1 liter.
Chlorine water (Cl_2,/H_2O)	Pass Cl_2 gas into distilled water until it is saturated. Prepare it when required because it spoils when kept for a long period.
Dichromate mixture	Dissolve 10 g of $Na_2Cr_2O_7$ or $K_2Cr_2O_7$ in water and add 25 ml of concentrated H_2SO_4. Make up the volume to 100 ml with distilled water.
2, 4-Dinitrophenyl-hydrazine	Dissolve 2 g of 2, 4-dinitrophenyl-hydrazine in 10 ml of concentrated H_2SO_4. Add this solution carefully to 200 ml of rectified spirit and filter after sometime.
Fehling's solution (For identification and estimation of reducing sugar)	Mix equal volumes of the following solutions A and B *Fehling's solution A or Copper sulphate solution:* Dissolve 34.66 g of crystalline copper sulphate in 500 ml of water containing a few drops of concentrated H_2SO_4. *Fehling's solution B or Alkaline tartarate solution:* Dissolve 173 g of sodium potassium tartarate or Rochelle salt ($KNaC_4H_4O_6.4H_2O$) and 50 g of NaOH pellets in 500 ml of water.

Contd...

Ferric chloride, ($FeCl_3$) solution	Prepare 1% solution, *i.e.*, dissolve 1 g of $FeCl_3$, in 100 ml of water.
Neutral $FeCl_3$ solution	When required add NH_4OH or Na_2CO_3 solution drop by drop to ordinary $FeCl_3$ solution until the precipitate formed, dissolves on shaking.
Mercuric chloride ($HgCl_2$) solution	Dissolve 68 g of $HgCl_2$ in distilled water and dilute to 1 liter.
Molisch's reagent or 15% Alcoholic α-naphthol solution (For carbohydrate and wool)	Dissolve 15 g of α-naphthol in 100 ml of rectified spirit (alcohol) or chloroform.
Phenyl hydrazine reagent	Dissolve 25 g of phenyl hydrazine in 100 ml of glacial acetic acid.
Picric acid reagent	Prepare a saturated solution of picric acid in benzene without heating. Keep it in a stoppered bottle. Dissolve 16 g of $KMnO_4$ in water and dilute to 100 ml.
Sodium bicarbonate ($NaHCO_3$) solution	Prepare its saturated solution.
Sliver nitrate (alcoholic)	Dissolve 10 g of $AgNO_3$ in 100 ml of rectified spirit or ethanol.
Sodium nitrite ($NaNO_2$) solution	Dissolve 2.5 g of $NaNO_2$ in 100 ml of cold water.
Sodium nitroprusside, $Na_2[Fe(CN)_5NO].2H_2O$ solution	Dissolve 1 g of the reagent in 10 ml of water. This solution should be freshly prepared whenever required.
Tollen's reagent (To be prepared when required)	Take 2 ml of $AgNO_3$ solution in a clean test-tube. Add to it 1-2 drops of NaOH solution when turbidity appears. Now add NH_4OH drop by drop until the turbidity to precipitate of AgOH disappears.

PREPARATIONS OF SOLUTIONS OF THE REAGENTS

Quality of reagents: Unless specified otherwise, chemicals or analytical reagent quality (A.R., G.R.) and distilled water (IS 1070 - 1960) shall be employed in preparing the solutions of reagents.

Acidic solutions

Acetic acid (IN): Dilute 57-40 ml of 17.4 N and 99.5% A.R. glacial acetic acid to 1 litre with distilled water in a measuring flask.

Hydrochloric Acid (IN): Dilute 86 ml of 11.6N and 36% concentrated A.R. hydrochloric acid to 1 litre with distilled water.

Sulfuric Acid (IN): Add 27.80 ml of 36 N and 98% sulfuric acid (conc.) to distilled water. After cooling make up the volume to 1 litre with distilled water.

Nitric Acid (IN): Dilute 61.8 ml of 16.2 N and 72% conc. A.R. nitric acid with distilled water to 1 litre.

Diluted acids

Acetic acid (5N) : Dilute 287 ml of A.R Glacial acetic acid (17.4 N, 99.5%) to 1 litre of distilled water.

Hydrochloric acid (5N): Dilute 430 ml of concentrated A.R. hydrochloric acid (11.6N., 36%) to 1 litre with distilled water.

Nitric acid (5N): Dilute 309.0 ml of conc A.R. Nitric and (16.2 N 72%) to 1 litre with distilled water.

Sulphuric acid (5N): Add slowly 139 ml of conc. A. R. Sulphuric acid (36N, 98%) of distilled water. After cooling make up the volume to 1 litre with distilled water.

Alkali solutions

Sodium hydroxide (1N): Dissolve 40 g of A.R. NaOH in 1 litre of Distilled water.

Potassium hydroxide (1N): Dissolve 56 g of A.R KOH pellets in 1 litre of distilled water.

Ammonium hydroxide (1N): Dissolve 66.6 ml of concentrated ammonia solution (28.4%, 15 N) to 1 litre of distilled water.

Ammonia solution (1 : 1): Mix equal volumes of concentrated ammonia solution (sp. gr. 0.88) and boiled out distilled water.

Buffer solutions

***Ammonium Chloride* : Ammonium hydroxide buffer solution (pH 10.0):** Dissolve 70 g of A.R. NH_4Cl and add 568 ml of conc. ammonia solution having specific gravity 0.88 to 0.90 stir and make up the volume to 1 litre with boiled out distilled water.

Phosphate buffer (pH 6.2-6.5): Transfer 24 g of anhydrous disodium hydrogen phosphate 46 g of anhydrous potassium hydrogen phosphate and 0.8 g of disodium salt of EDTA in a 1 litre standard measuring flask and make up the volume to 1 litre with boiled out distilled water.

Phosphate buffer (pH 7.2): Dissolve 8.50 g KH_2PO_4 $7H_2O$ and 21.75 g Na_2HPO_4 and 1.70 g NH_4Cl in boiled out distilled water and make up the volume to 1 litre in a measuring flask.

Calcium precipitating buffer (pH 8.0): Dissolve 6.0 g of A.R.. $(NH_4)_2$ C_2O_4 (ammonium oxalate) in about 100 ml of boiled-out distilled water. Add 144 g of A.R. NH_4Cl, 13 ml of concentrated ammonia solution and make up the volume to 1 litre with distilled water.

Ammonium acetate buffer solution (pH 5.0): Dissolve 250 g of ammonium acetate in 150 ml of distilled water. Add 700 ml of glacial acetic acid.

Indicator solution

Eriochrome Black T: Dissolve 1 g of pure solid dye in 75 ml of triethanol amine and 25 ml of distilled ethanol. This reagent is stable for several months.

or

Dissolve 0.5 g of the dye stuff in 100 ml rectified spirit or methanol. This solution can be used for atleast a month.

Calcon indicator: Dissolve 0.2 g of dyestuff in 50 ml of methanol.

Murexide indicator: Dissolve 10 mg of murexide with 490 mg of sodium chloride and use 2 mg of this solid mixture for each titration.

For indicator solution prepare saturated solution of murexide in distilled water.

Phenolphthalein: Dissolve 1 gm of phenolphthalein in 100 ml of distilled ethyl alcohol and add 100 ml of distilled water with constant stirring, filter, if necessary.

Potassium Chromate: Dissolve 5.0 gm of A. R. potassium chromate in 100 ml of distilled water. Use 1 ml of the indicator for each titration.

Starch: Prepare a paste of 1 g of powdered starch (soluble A.R.) with distilled water and add to this paste 100 ml of boiled water with constant stirring. Boil for about five minutes and then cool.

Ferroin: Dissolve 3.7125 gm of O-phenanthroline monohydrate with 1.6375 $FeSO_4 \cdot 7H_2O$ in distilled water and make up the volume with distilled water to 250 ml. The indicator solution itself may be purchased from the market.

Methyl Orange: Dissolve 50 mg of free acid form of methyl orange solid in distilled water and make the volume to 100 ml with distilled water. Filter the solution if necessary.

For sodium salt form of methyl orange solid, dissolve 50 mg in 100 ml of distilled water. Add 1 ml of 0.1 M HCl and filter if necessary.

Methyl red: Dissolve 0.25 g of free acid form of methyl red solid in 250 ml of hot distilled water. Cool and filter if necessary.

Dissolve 0.1 g in 60 ml of distilled alcohol and add 40 ml of distilled water.

Methylene blue indicator: Dissolve 1 gm of powder indicator in distilled water and make up the volume to 500 ml.

Litmus solution (blue): Dissolve one gm of solid in 100 ml of distilled water.

Litmus solution (red): To the above blue litmus solution add a drop or two of dil. HCl in order to change its colour to red.

Ferric Alum: Dissolve 25 g of $Fe_2(SO_4)_3\ (NH_4)_2\ SO_4 \cdot 24H_2O$ in 100 ml of hot distilled water. Add 10 ml of A.R. Nitric acid and boil until yellow coloured solution is obtained. Cool and filter the solution.

Potassium ferricyanide: Dissolve 55 gms of substance in one litre of distilled water (0.5 N).

Methyl violet indicator: Dissolve one gm of indicator powder in 25 ml 95% of alcohol and make up the volume to 100 ml with distilled water.

Solutions of approximate strength/Standard solution

Standard EDTA solution (M/100 or (N/50): Weigh 3.723 g of A.R. disodium dihydrogen ethylene diamine tetra acetic acid ($Na_2\ H_2C_{10}$-$H_{12}\ O_8N_2 2H_2O$) having molecular weight 372.25 (eq. wt. 186.125) before weighing, it is dried in an air oven at 75°C for 90 minutes and cooled. Dissolve it in distilled water in a measuring flask and make up the volume to 1 litre with distilled water.

Hard water (1 ml = 1 mg $CaCO_3$): Add slowly a small amount of dilute hydrochloric acid, through a funnel, to 1 g of anhydrous $CaCO_3$ (A.R. grade) taken in a conical flask. Boil gently to remove CO_2. Heat to dryness on a water bath. Dissolve in boiled out distilled water and dilute to 1 litre.

Standard NaCl solution (N/10): Dissolve 5.845 gm of pure sodium chloride (NaCl) in distilled water and make up the volume to 1 litre.

Silver nitrate solution (N/50): Dissolve 3.3978 g of silver nitrate solid (having mol. wt. 169.89) in 1 litre of distilled water.

Standard Potassium dichromate solution (N/10): Dissolve 4.904 g of pure potassium dichromate ($K_2Cr_2O_7$) in distilled water and dilute to 1000 ml.

Sodium Thiosulphate (N/10): Dissolve 24.82 g of A.R. $Na_2S_2O_3 \cdot 5H_2O$ (Mol. wt. 248.21 eq. wt 248.21) in 1 litre of boiled out distilled water.

Sodium sulfite (0.10 N): Dissolve 6.3 g Na_2SO_3 in distilled water and make up the volume to 1 litre. Use the fresh solution whenever required as the stability of the solution is poor.

Sodium Nitroprusside: Dissolve 3 g of substance in one litre of distilled water.

Sodium hydrogen phosphate: Dissolve 120 g Na_2HPO_4 $12H_2O$ in one litre of distilled water (1 N)

Potassium hydrogen phthalate solution (N/10): Weigh 20.410 g of potassium hydrogen phthalate (H K C_8 H_4 O_4) dissolve in boiled distilled water and make up the volume to 1 litre with distilled water.

Potassium permanganate (N/10): Dissolve 3.1607 g of A.R $KMnO_4$ having mol.wt 158.037 and eq. wt. 31.607 in distilled water. Boil for 1 hour, cool, filter and make up the volume to 1 litre with distilled water.

Potassium Iodate (N/100): Dissolve 3.567 g of A.R. KIO_3 having mol. wt. 214 and eq. wt 35.66 in distilled water and make up the volume to 1 litre.

Standard copper sulfate (N/10): Dissolve 24.96 g of A.R. $CuSO_4$ $5H_2O$ having mol. wt. 249.6 in distilled water and make up the volume to 1 litre.

Ferric Chloride (0.1M): Dissolve 27.05 g of $FeCl_36H_2O$ having mol. wt. 270.5 in distilled water with 50 ml of dil. HCl. Make up the volume to 1 litre with distilled water.

Potassium iodide (0.1 M): Dissolve 16.60 g of iodate free, A.R.KI in distilled water and make up the volume to 1 litre.

Stannous chloride (0.1M): Dissolve 2.257g of $SnCl_2.2H_2O$ having mol. wt. 225.7 in 5 ml of conc. hydrochloric acid and make up the volume to 1 litre with distilled water.

Sodium bisulfite ($NaHSo_3$)(N/10): Dissolve 5.20 g of $NaHSO_3$ having mol.wt. 104 and eq. wt. 52.0 in distilled water containing 50 ml of 2N H_2SO_4 and make up the volume to 1 litre with distilled water.

Oxalic acid (N/10): Dissolve 6.303 g of A.R. $(COOH)_2$. $2H_2O$ having mol. wt. 126.068 and eq.wt. 63.034 in distilled water and make up the volume to 1 litre. This is used to standardise the $KMnO_4$ solution.

Ceric Ammonium Nitrate solution: Dissolve 40 g of the substance to 100 ml of hot. dil. HNO_3 (2N).

Barium cloride: Dissolve 61 g of substance ($BaCl_2.2H_2O$) in one litre of distilled water (0.5N)

Ammonium thiocyanate: Dissolve 38 g of substance in one litre of distilled water (0.5N).

Ammonium Oxalate: Dissolve 35 g of substance in one litre of distilled water (0.5N).

Ammonium Nitrate: Dissolve 80 g of substance in one litre of distilled water (1N).

Calcium hydroxide (Lime water): Dissolve 1.7 g of CaO or 2-3 g of $Ca(OH)_2$ in one litre of distilled water. Decant the solution after long standing (0.04N).

Potassium Nitrate: (0.1M) Dissolve 9.1 g of Pure KNO_3 (Mol. wt. 91) in 1 litre of distilled water.

Ethyl acetate (M/10): Add 9.8 ml of pure ethyl acetate (Mol. wt. 88.1, SP. gr. 0.9005 g/ml at 20°C) to distilled water and make the volume to 1 litre.

Ferrous Ammonium Sulphate (N/4): Dissolve 49.0 g of A.R. FAS ($FeSO_4(NH_4)2SO_4.6H_2O$) in boiled out distilled water containing 10 ml of conc. H_2SO_4 and dilute to 500 ml. Standardise with N/4 $K_2Cr_2O_7$ solution.

Potassium Ferrocyanide: Dissolve 53 g of substance in one litre of distilled water (0.5N).

Alcoholic Potassium Hydroxide (N/100): Dissolve 5.6 g of A.R. KOH (Mol. wt. 56.0, eq wt. 56) pellets in 1 litre of 95% alcohol. Mix throughly and let stand undisturbed, for any carbonate to settle down. Decant the clear supernatant solution and use and standardise with N/10 oxalic acid.

Bismuth Chloride (0.1M): Dissolve 31.5 g of $BiCl_3$ (Mol. wt. 315.5) in the minimum volume of 5NHCl and dilute to 1 litre. Add dropwise, with stirring, sufficient hydrochloric acid to remove any precipitate.

Ferric Nitrate (0.1M): Dissolve 40 g of $Fe(NO_3)_3.9H_2O$ (Mol. wt. 404) in 1 litre of distilled water containing 50 ml of dilute nitric acid.

Ferric Sulphate (N/2): Dissolve 14 g of $FeSO_4.7H_2O$ (Mol. wt. 278, Eq. wt. 278) in 100 ml of distilled water containing 10 ml of dilute Sulphuric acid.

Ferrous ammonium sulphate (N/10): Dissolve 39.214 g of $Fe(NH_4)_2.(SO_4)_2 6H_2O$ having mol. wt. 392.14 in distilled water. Add 20 ml of conc. H_2SO_4 and make up the volume to 1 litre.

Sodium carbonate (N/10): Dissolve 5.30 g of A.R. Na_2CO_3 (Mol. wt. 106, eq. wt. 53) dried at 250°C in boiled out distilled water and dilute to 1 litre with distilled water.

Oxalic acid (0.6M): Dissolve 75.6 g of oxalic acid in 1 litre of distilled water.

Manganese sulfate: Dissolve 480 g of $MnSO_4.4H_2O$ in distilled water and make up the volume to 1 litre.

Mangnesium sulfate (for BOD): Dissolve 22.50 g of $MgSO_4.7H_2O$ in distilled water and make up the volume to 1 litre.

Calcium chloride (for BOD): Dissolve 27.5 g $CaCl_2$ in distilled water and make up the volume to 1 litre.

Bromine Solution (Saturated): Shake 11 ml of liquid bromine with 1 litre of distilled water.

Ammonium Acetate Solution: Dissolve 200 g A.R. Sodium acetate in 800 ml distilled water.

1,10-Phenanthroline Solution: Dissolve 100 g of 1, 10-Phenanthroline Monohydrate $Cl_2H_8N_2.H_2O$ in 100 ml of distilled water. Warm slightly or add 2 drops of conc. HCl. 1 ml of this solution can chelate 100 mg of iron.

Hydroxyl amine Hydrochloride: Dissolve 10 g of $NH_2OH.HCl$ in 100 ml distilled water.

Stock Iron Solution: Dissolve 1.404 g of $Fe(NH_4)_2.(SO_4)_2.6H_2O$ in 50 ml distilled water to which 20 ml conc. H_2SO_4 is added earlier. Add 0.1N $KMnO_4$ solution dropwise till a faint pink colour persists. Dilute to 1 litre with distilled water.

Alkaline iodide-azide reagent: Dissolve 700 g of KOH and 135 g NaI (500 g of NaOH and 150 g KI) in distilled water and dilute to 1 litre. Add 10 g sodium azide dissolved in 40 ml of distilled water.

Iodine solution: Dissolve 20 g of KI and 13 g of iodine in 30 ml of distilled water and make up the solution to one litre. (0.1N)

Sucrose (10%): Dissolve 10 g of granulated sufar in 100 ml distilled water.

Alkaline Potassium Iodide: Dissolve 700 g of KOH pellets and 150 g of A.R. KI in distilled water and dilute to l litre.

Ferric Sulphate (0.5%): Dissolve 5 g of Ferric Sulphate (M. wt. 400) in 1 litre of distilled water.

Mercuric Chloride(Saturated): Dissolve 80 g of $HgCl_2$ in 1 litre of hot distilled water. Cool to room temperature and filter.

Potassium Oxalate (2%): Dissolve 2 g of potassium oxalate ($K_2C_2O_4.H_2O$) in 100 ml of distilled water.

LABORATORY REAGENTS

1. Concentrated Acids

Name and Formula of Acid	Formula Weight	Equivalent Weight	Weight %	Specific Gravity	Aprox. Strength
Acetic acid (CH_3COOH) Glacial	60.05	60.05	99-100 or 99.5	1.05	17.4 N or 17.4M
Hydrochloric acid (HCl)	36.46	36.46	36	1.18	11.6 N or 11.6 M
Hydrobromic acid (HBr)	80.92	80.92	47.8	1.49	9 N or 9 M
Hydrofluoric acid (HF)	20.01	20.01	46	1.5	26.5 Nor 26.5 M
Hydriodic acid (HI)	127.91	127.91	47.475	1.50	5.5 N or 5.5 M
Nitric acid (HNO_3)	63.01	63.01	69.5 or 70	1.42	16 N or 5.5 M
Phosphoric acid (H_3PO_4)	98.00		85	1.69	40.5 N or 13.5 M
Perchloric acid ($HClO_4$)	100.46		70	1.66	9 N or 9 M
Sulphuric acid (H_2SO_4)	98.08	49.0	96	1.84	36 N or 18.0 M

2. Bases

Acid	Equivalent Weight	Aporox. Strength	Preparation of 1 liter Solution
Concentrated ammonium hydroxide (NH_4OH)	35.05	1.51 N or 58.6% by weight of NH_4CH or 28.30% NH_3.	Highest concentration with specific gravity 0.90.
Dilute ammonium hydroxide (NH_4OH)	35.05	5 N (56.5 ml concentrated alkali per liter will give N solution and 166 ml/500 will give 5 N solution).	Dilute 332 ml of concentrated ammonia (specific gravity 0.90 with distilled water by making it to 1 liter.
Sodium hydroxide (NaOH)	40.00	5 N	Dissolve 220 g of caustic soda pellets in distilled water and dilute to 1 liter.
Lime water, $Ca(OH)_2$	= 37.05	0.04 N	Shake 10 g of lime (sufficient to saturate 1 liter of water) with 1 liter of water and allow it to stand for some-time. Filter and keep it in a stoppered bottle.

Viva Questions

1. Write the unit of hardness?
2. What is structure of EDTA? Write the abbreviation for EDTA.
3. Why is hardness of water generally expressed as $CaCO_3$ equivalent?
4. What are the end-points in EDTA titrant?
5. What is the cause of hardness of water?
6. Name the indicator used in the titration of Ca^{2+} and EDTA.
7. What are compleximetric titrations?
8. What is meant by hard water and its hardness?
9. How is this optimum pH attained?
10. What are metallochromic indicators?
11. How is the color of the indicator affected by the pH?
12. What are the harmful effects of hard water?
13. Name a metallochromic indicator other than EBT.
14. Define acidimetry and alkalimetry?
15. Which ions could the alkalinity of water?
16. How can you determine the extent of alkalinity present in water sample?
17. Which indicator is used in the titration of strong acid (HCl) and strong base (NaOH)?
18. What is indicator. How does it function?
19. What is pH range of phenolphthalein and methyl orange?
20. What are the demerits of the highly alkaline water?
21. What is 'available chlorine'?
22. What is meant by iodometric titration?
23. Give the formula of bleaching powder.
24. Write the chemical reactions involved in the experiment.
25. Why is only freshly prepared starch used as indicator?
26. What is the color change at the end point?
27. Why does the blue color appear on the addition of starch?
28. Why should the starch indicator be added near the end point?
29. Why is an excess of KI always used?
30. What is a conductometric titration ?
31. What is relationship between conductivity and specific conductivity ?
32. What are the factors affecting conductivity ?
33. Which cation and anion has highest conductivity ?

34. How is the end point determined in conductometric titration ?
35. What are the applications of conductometric titrations ?
36. What is the effect of dilution on the molar conductivity of a weak electrolyte ?
37. What is the effect of temperature on molar conductivity ?
38. Why does the equivalent conducicitivity of electrolyte increases with dilution ?
39. What is the effect of dilution on *p*H of an acidic solution?
40. What chemicals would you use to make a buffer of *p*H (i) 5, (ii) 10?
41. What is a combined electrode? What is the mechanism of its working?
42. What is the desirable *p*H range for drinking water?
43. What are calomel and glass electrodes?
44 Define emf of a cell.
45. Glass electrode is preferred to quinhydrone electrode in measuring pH of a solution. Give reason?
46. Why Hydrogen electrode not generally used in pH measurements?
47. What chemicals would you use to make a buffer of pH (i) 5 (ii) 10 ?
48. What is calomel electrode ?
49. What is the potential of hydrogen electrode ?
50. What are the units of dissociation constant?
51. What happens to the degree of dissociation of an electrolyte with changes in concentration and temperature?
52. How is stirring useful during precipitation?
53. Define the term 'digestion'
54. Why the precipitate is washed?
55. In estimation of lime by $KMnO_4$, what indicator you have employed?
56. In lime estimation how calcium oxalate precipitate is converted into $H_2C_2O_4$?
57. What is lime and how much percentage of Calcium generally present in lime?
58. What are alloys? Is bronze an alloy? What is the typical composition of bronze?
59. What makes steel "stainless"?
60. Write equations for all the reactions which are taking place during the dissolution of brass.
61. Can you some other reducing agent in place of iodide ion to reduce Cu^{++} to Cu^{+} for the determination of copper in brass? Give reasons for your answer.
62. Electrode potentials depend upon the concentration of a particular ion in solution. Can you calculate the minimum concentration of copper which will be reduced by iodide ion.
63. What is pyrolusite ?
64. In estimated of pyrolusite what indicator is employed ?
65. What is the principle involved in estimation of pyrolusite ?
66. What is the element present in the pyrolusite ?
67. What is Beer's Law ?

68. What is Lambert's Law ?
69. Give combined law of "Beers-Lamberts"
70. What is the importance of colorimetric estimation ?
71. What is calibration curve ?
72. What happened in the reaction of Fe^{3+} and KCN ?
73. How will you differentiate qualitatively between aspirin and salicylic acid? Suggest a color test?
74. Write the molecular formula of aspirin?
75. Aspirin can also be called as?
76. How will you ascertain the aliphatic or aromatic nature of the given compound?
77. Why do you fuse sodium with organic compound?
78. Explain Lassaigne's test.
79. Why is metallic sodium kept in kerosene oil?
80. Can potassium be used instead of sodium?
81. What is the nature of sodium extract?
82. Why should ferrous sulphate solution be fresh and saturated?
83. Sometimes a blood red coloration is produced on addition of $FeCl_3$ in sodium extract, why?
84. What is Baeyer's reagent?
85. If a compound turns blue litmus to red what will you guess?
86. How can you detect the presence of alcoholic group?
87. What is the difference between alcoholic group and phenolic group?
88. How can you detect the carbonyl group?
89. How can you distinguish between aldehydes and ketones?
90. How can you detect the presence of carbohydrates?
91. Which one is more acidic, phenol or carboxylic acid?
92. How can you identify a primary amine?
93. How can you detect a $-NO_2$ group?
94. What are the units of rate of constant of:

    (i) zero order (ii) I order (iii) II order (iv) III order (v) Half order
95. How is rate constant related to concentration of reactants?
96. Why is a reaction speed up in presence of catalyst?
97. Give an example of Pseudo-unimolecular reactions?
98. Saponification reaction comes under which order?
99. The activation energy for a chemical reaction depends upon which factor?
100. Give an expression for Arrhenius equation showing the effect of temperature on the rate constant is $(T_2 > T_1)$
101. On what factor specific rate constant of a 1st order reaction depends

102. What is the order of the reaction which obeys the expression $H_2 - \frac{1}{k_a}$

103. What are the dimensions of viscosity?

105. Give the units of viscosity in SI and CGS system.

106. What is the effect of temperatur on viscosity?

107. Discuss the factor on which viscosity of a liquid depends.

108. What are the applications of viscosity measurements?

109. Define the terms adsorption and absorption.

110. Define the term adsorbent and adsorbate.

111. What do you mean by adsorption isotherm?

112. What do you mean by positive adsorption and negative adsorption?

113. What are the various types of adsorption isotherms?

114. Give the application of adsorption.

115. On what factors adsorption depend?

116. What is the effect of temperature on physical and chemical adsorption?

117. What is the cause of adsorption?

118. How do residual forces arise on the surface of a solid adsorbent?

119. Distinguish between physical and chemical adsorption.

120. Write balanced equations for all the chemical reactions in the above experiment.

121. Why does the potassium dichromate solution make the print sharper?

122. $E = h\gamma = hc\gamma' = hc/\lambda$. Explain all the terms. What are the units of h?